1988

Focus On

Phytochemical Pesticides

Volume I

The Neem Tree

Editor

Martin Jacobson

U.S. Department of Agriculture (Retired)
Silver Spring, Maryland

CRC Press
Taylor & Francis Group
Boca Raton London New York

CRC Press is an imprint of the
Taylor & Francis Group, an **informa** business

First published 1989 by CRC Press
Taylor & Francis Group
6000 Broken Sound Parkway NW, Suite 300
Boca Raton, FL 33487-2742

Reissued 2018 by CRC Press

© 1989 by CRC Press, Inc.
CRC Press is an imprint of Taylor & Francis Group, an Informa business

No claim to original U.S. Government works

Publisher's Note
The publisher has gone to great lengths to ensure the quality of this reprint but points out that some imperfections in the original copies may be apparent.

Disclaimer
The publisher has made every effort to trace copyright holders and welcomes correspondence from those they have been unable to contact.

ISBN 13: 978-1-315-89295-5 (hbk)
ISBN 13: 978-1-351-07205-2 (ebk)

Visit the Taylor & Francis Web site at http://www.taylorandfrancis.com and the
CRC Press Web site at http://www.crcpress.com

FOCUS ON PHYTOCHEMICAL PESTICIDES

SERIES INTRODUCTION

There is ample evidence to show that the plant kingdom is a vast storehouse of chemical substances manufactured and used by plants for defense from attack by insects, bacteria, and viruses. Many of these plants, as well as their chemical components, have been used since ancient times to prevent and treat diseases occurring in higher animals, including humans. The scientific and pseudoscientific literature is replete with such reports, some of which have been confirmed by laboratory and clinical trials, and yet others which have been shown to be worthless for these purposes.

The CRC Series *FOCUS ON PHYTOCHEMICAL PESTICIDES* is envisioned as a comprehensive series of state-of-the-art volumes covering all aspects of plant use as crop protectants from attack by insects, diseases, fungi, nematodes, and predatory wildlife such as coyotes, wolves, and rodents. Coverage will not, however, be limited to protection of crops but will also include the use of plants to protect humans and farm animals from diseases transmitted through mollusks, fungi, and viruses. Each volume in this series will contain detailed compilations contributed by known authorities in the particular field. The high quality of the information contained in these volumes will be assured with the aid of an Advisory Board composed of worldwide authorities in the phytochemical pesticide field.

M. Jacobson
Editor-in-Chief

PREFACE
Volume I: The Neem Tree

Approximately one third of the world food crop is damaged or destroyed by insect pests during growth, harvest, and storage. Losses are considerably higher in many underdeveloped countries of Asia and Africa. The monetary loss due to feeding by larvae and adults of pest insects amounts to billions of dollars each year. Furthermore, the comfort and well being of humans and beneficial animals are affected directly by household and environmental pests such as lice, ants, roaches, ticks, wasps, and mosquitoes, some of which are disease carriers and transmitters. Many of the synthetic pesticides previously used for insect control have been banned or their use seriously curtailed because of concern about health and environmental effects. Also, the adaptability of insects threatens to undermine the effectiveness of existing pesticides. It is therefore imperative that safe, biodegradable substitutes for the synthetic pesticides be discovered.

Over the years, a wealth of literature has accumulated in scientific journals, books, and other reports on the effectiveness of plants as insect feeding deterrents, repellents, toxicants, and disruptants of insect growth and development. Heading the list of effective plants, from the standpoints of number of pest species affected, high activity, availability, safety, and resistance to predators, is the subtropical neem tree, *Azadirachta indica* A. Juss, a hardy member of the plant family Meliaceae. It is therefore fitting that the first volume in the series *Focus On Phytochemical Pesticides* should be devoted exclusively to this "wonder tree".

Based on the wealth of scientific records, both oral and printed, on the effectiveness of all parts of this tree against countless species of insects, nematodes, bacteria, and viruses, the First International Neem Conference was held June 16—18, 1980, at Rottach-Egern, West Germany, with more than 40 scientists (chemists, entomologists, botanists, physiologists, and zoologists) from 4 continents in attendance. At this Conference, an informal steering committee was appointed that quickly became known as the "Neem Mafia", consisting of Professors K. R. S. Ascher (Volcani Center, Bet Dagan, Israel), E. D. Morgan (University of Keele, U. K.), H. Rembold (Max-Planck-Institute of Biochemistry, Munich, West Germany), R. C. Saxena (International Rice Research Institute, Manila, Philippines), H. Schmutterer (Justus Liebig University, Giessen, West Germany), L. M. Schoonhoven (Agricultural University, Wageningen, The Netherlands), and me. This Conference was so successful in stimulating worldwide interest and activity in the efficacy of the neem tree that it was followed by the Second International Neem Conference, held May 25—28, 1983, in Rauischholzhausen, West Germany (105 participants), and the Third International Neem Conference, held July 10—17, 1986, in Nairobi, Kenya (64 participants).

We express our heartfelt thanks to the Gesellschaft für Technische Zusammenarbeit (GTZ), Eschborn, West Germany, for sponsoring and providing generous funding of all three Conferences, making possible the attendance and participation of numerous scientists from Asia and Africa, where the neem tree is naturally endemic and widely used as a pesticide. We are also grateful to Drs. B. S. Parmar and R. P. Singh, Indian Agricultural Research Institute (IARI), New Delhi, India, for their continuing efforts to keep neem scientists worldwide informed of progress through quarterly publication and distribution of the *Neem Newsletter,* which began in 1984.

Cooperative research between botanists, chemists, and entomologists has very recently resulted in the commercialization of neem formulations as insect control agents in Asia, Europe, and the U. S., as well as the successful cultivation of the tree in many areas where it had not previously existed.

M. Jacobson

THE EDITOR

Martin Jacobson received his B.S. degree in chemistry from the City University of New York in 1940. He then accepted an offer as a chemist with the Industrial Hygiene Division of the National Institutes of Health in Bethesda, MD. In 1942, he transferred to the Bureau of Entomology and Plant Quarantine of the U.S. Department of Agriculture (USDA) Agricultural Research Center, Beltsville, MD, as a research chemist to isolate, identify, and synthesize phytochemical pesticides, insect hormones, and insect sex pheromones. During this period he pursued evening graduate studies in chemistry and microbiology at George Washington University, Washington, D.C. He also served as a part-time Research Associate in Chemistry at that University during the period 1944 to 1948.

From 1964 to 1972, Mr. Jacobson was an Investigations Leader with the Entomology Research Division at Beltsville, Chief of the Biologically Active Natural Products Laboratory from 1973 to 1985, and Research Leader (Plant Investigations) with the Insect Chemical Ecology Laboratory until his retirement from Federal Service in 1986. He is currently an agricultural consultant in private practice in Silver Spring, MD.

During his long career with the USDA, Mr. Jacobson spent several weeks in 1971 as a Visiting Scientist teaching a graduate course on insect pheromones and hormones in the Department of Chemistry, University of Idaho, Moscow. He was invited to organize numerous symposia and speak at national and international scientific meetings in the U.S., Europe, Asia, and Africa, in the field of pesticides and sex pheromones occurring naturally in plants and insects, respectively. His awards include the Hillebrand Prize of the Chemical Society of Washington in 1971; USDA Certificates of Merit and cash awards for research in 1965, 1967, and 1968; the McGregory Lecture Award in Chemistry at Colgate University (Syracuse, NY); two bronze medals for excellence in research at the 3rd International Congress of Pesticide Chemistry, Helsinki, Finland in 1974; USDA Director's Award on Natural Products research in 1981; and an Inventor's Incentive Award for commercialization of a boll weevil deterrent in 1983.

Mr. Jacobson has been a member of the American Chemical Society, Chemical Society of Washington, Pesticide Science Society of Washington, American Association for the Advancement of Science, New York Academy of Sciences, and a Fellow of the Washington Academy of Sciences. He is the author or coauthor of more than 300 scientific reports in numerous journals, the author of four books (*Insect Sex Attractants,* John Wiley & Sons, 1965; *Insect Sex Pheromones,* Academic Press, 1972; Insecticides From Plants. A Review of the Literature, 1941—1953, USDA Handbook 154, 1958; Insecticides From Plants. A Review of the Literature, 1954—1971, USDA Handbook 461, 1975), and editor of the book *Naturally Occurring Insecticides,* Marcel Dekker, 1971. He also holds six U.S. Patents on naturally occurring insecticides.

TABLE OF CONTENTS

Cultivation and Propagation of the Neem Tree

Michael D. Benge
Office of Forestry, Environment, and Natural Resources
Bureau for Science and Technology
Agency for International Development
Washington, D.C.

TABLE OF CONTENTS

I. DISTRIBUTION

Neem is thought to have originated in Burma and is common throughout the open scrub forest in the dry zone and on the Siwalik hills; and if it is native to India, it occurs naturally only in Karnatak and in parts of the Deccan Peninsula.[1] In its native environments, neem is generally found growing in mixed forests, and associated with other broadleaf species, such as *Acacia* sp. and *Dalbergia sissoo*. It grows in tropical to subtropical regions; semiarid to wet tropical regions; and from sea level to over 610 m (2000 ft). It is cultivated throughout India, and in many places has become wild. (The fruit is not toxic to birds and bats and it is reported that they are mainly responsible for the spread of wild neem.) It is now found in many places outside its native distribution to the sub-Himalayan track, including the northern part of Uttar Pradesh at 610 m (2000 ft) and the southern part of Kashmir at 670 m (2200 ft).[2]

However, according to Ahmed and Grainge,[3] neem is native to the Indo-Pakistan subcontinent, while others attribute its nativity to the dry forest areas of India, Pakistan, Sri Lanka, Malaysia, Indonesia, Thailand, and Burma.[4] Nevertheless, neem is found throughout Pakistan, Sri Lanka, southern Malaysia, Indonesia, Thailand, and the northern plains of Yemen and has been recently introduced into Saudi Arabia. The Philippines has also begun widescale plantings of neem for fuelwood and pesticide production. It is known that 19th century immigrants carried the tree from the Indo-Pakistan region to Fiji, and it has now spread to other islands in the South Pacific.

East Indian immigrants introduced neem to Mauritius, and it is thought that they took it to a number of African countries. It is now widely cultivated on the African continent in Ethiopia, Somalia, Kenya, Tanzania, Mozambique, Mauritania, Togo, Ivory Coast, Cameroon, Nigeria, Guinea, Ghana, Gambia, Sudan, Benin, Mali, Niger, Burkina Faso, Chad, and Senegal, particularly in rainfall-deficient regions.

Neem was also introduced into several Caribbean nations by East Indian immigrants, and is now propagated by Indian communities as a medicinal plant in Trinidad and Tobago, Jamaica, Suriname, Guyana, and Barbados.[5] Neem plantings are also abundant in Guatemala, Bolivia, Ecuador, Honduras, Argentina, Brazil, Cuba, Nicaragua, Dominican Republic, St. Lucia, and Antigua and large numbers are now being planted in Haiti. In addition to the ongoing experimental cultivation of neem in Puerto Rico and the U.S. Virgin Islands, plantings in southern Florida are thriving, and the field cultivation of neem in Oklahoma, southern California, and Arizona has begun.[3,6-9]

II. ADAPTATION

Neem grows well from sea level to over 670 m (2000 ft), and can be established in hot and dry regions without irrigation. Neem thrives under subhumid to semi-arid conditions and can be established in areas with an annual rainfall of 450 to 750 mm (18 to 30 in.). Optimum growth is obtained in higher rainfall areas (1150 mm); 130 mm/year is sufficient for survival,[4] but it needs 450 mm to grow successfully. It grows where maximum shade temperature may be as high as 49°C (120°F), but it does not stand excessive cold. Neem is frost-tender, especially in the seedling and sapling stages, but it is grown in frost zones of the sub-Himalayan tract by protecting seedlings during the winter with screens.[7] Fire often kills it outright.[2,8,9] According to Gorse,[10] neem is not a very sociable tree, does not grow well in pure stands (plantations in Africa often die out in 3 to 10 years), and is very competitive for water and soil nutrients; thus it does not grow well on marginal soils.

Neem seems to grow best in deep sandy soils that are well drained, but can grow in practically all sorts of soil; it thrives on black cotton soil in India and does not do badly even on clay. It does better than most species in dry localities, on sandy, stony shallow soils with a waterless subsoil or in places where there is a hard calcareous or claypan (hardpan) not far from the surface. However, neem will grow much better if this hardpan is broken up before planting. Occasionally, neem will initially grow well on soils that appear to be sandy, but quickly die out when roots hit a deep layer of dense clay. Neem can grow even on saline and on alkaline "usar" soils:[11] however, neem has been reported to be susceptible to moderate salinity in the Sudan.[12] Thus within the species, some provenances must be more genetically tolerant to salinity than others. Fishwick[13] observes that the best neem growth is found on sites with a soil pH of between 6.2 and 7+.

Neem does not tolerate waterlogged soils, and does not do well on soils with impeded drainage and on soils subject to inundation. Growth is not good on poorly drained soils, because the taproot tends to rot and the trees gradually die. In Nigeria, De Jussieu[14] reported that the best neem groves grew where the water table was 1.5 to 1.75 m down; at 2.5 m, the groves were only mediocre; and at 8 m, the groves died out. In his report, the method of planting the neem trees was not given, and in many places in Africa it was commonplace to establish plantings by stump cuttings, whereby the taproot was pruned. When this is done, a normal length taproot will not regenerate, and the tree develops extensive lateral root systems with only pseudo-taproots developing that do not penetrate deep into the subsoil. In some soils, the water table markedly rises during the rainy season, and if this rise persists, the roots of neem trees may smother.

Neem does not grow well in soils with high proportions of very fine sand or silt or finely divided mica; the yellowing of leaves is often followed by death.[8] This may be a result of nutrient deficiency. Research by Zech[15] determined that zinc and potassium deficiencies reduce neem growth evidenced by chlorosis of leaf tips and leaf margins, particularly on the older leaves: the first symptom of zinc deficiency is yellow coloration of the intercostal areas leading to complete breakdown of the chlorophyll. The shoots exude much resin with shedding of the older leaves. With potassium deficiency, leaf tip and marginal chlorosis and necrosis result.

In Nigeria, where systematic tests have been conducted, it was observed that very successful neem groves could be found in soils having a high clay content (67%), while trees died out on soils with a high sand content (83%).[14]

Neem has been planted on plantation scale in Nigeria since 1936. Most of the forests in the Sokoto region are of this species. The introduction of neem to Sokoto was cited as the greatest boon of the century. The tree grew quickly and met the local demand for firewood and poles for house construction and fence posts, in addition to providing welcome shade in towns and villages.[16] Gorse[10] states that neem is often the preferred species for planting in semiarid areas because animals do not readily browse it, and when they do, mortality is low and neem recovers rapidly. However, Welle[17] reports severe stunting of neem girdled from bark feeding by goats.

III. SILVICULTURAL AND PLANTATION CHARACTERISTICS

In India, neem is generally found in mixed forests. So far, little is known about the behavior of neem in plant communities and its ecological potential and limitations. Individual neem trees are reported to live for 200 years.[2] It seems to do much better on an isolated basis than in full groves, and grows beautifully along roadsides, or as isolated shade trees. It also does well in mixed species plantings and in relatively well-spaced rows (approximately 3 m apart) along contour line ditches.[14] Michel-Kim and

Brandt[18] report that experience in India and Africa indicates that neem may not be suited for monoculture. This contention is supported by Gorse[10] who states that neem is not a sociable tree and plantations often die after 3 to 10 years, especially on poorer sites. He contends that neem must be pruned or it will die back, which can be avoided by pollarding the tree. However, many report neem growing well in monoculture in plantations.[9,13,19,20] One explanation for this variance is that neem is very demanding on both water and mineral nutrients, and on soils where nutrients are limited neem will not do well in monocultural plantations.

Neem grows better with shade in its early stage of growth, but demands light as it matures. Therefore, it has a great capacity for pushing its way through thorny scrub in its youth. Welle[17] reports neem growing exceedingly well in plantations in Haiti where *Leucaena leucocephala* is interplanted as a nurse crop, with trees having a better form (the elephant-foot stump type growth with a large taper is reduced). Radwanski and Wickens[8] report that transplants are likely to be seriously injured by developing leaf spot (clorosis) as a result of insolation (grown in full sunlight), thus neem does best in its earlier years when planted with a nurse crop. However once mature, neem does best in full sunlight, and does not do well as an understory.

IV. GROWTH

Neem is reported to grow relatively fast, but varies greatly depending upon its environment, site characteristics, and the genetic capability of the plant material. Slower growth results at higher elevations, at colder temperatures, and on drier sites. Radwanski[7] reports that 66% of the total growth of the tree will occur in the first 3 years, during which it will reach a height of 4 to 7 m; it will reach 5 to 11 m in the following 5 years. Seedlings show moderate development, ordinarily reaching a height of 10 to 20 cm (4 to 8 in.) by the end of the first year. As a rule, trees put on a mean-annual girth increment of 2 to 3 cm (0.9 to 1.2 in.), though more rapid growth is obtained under more favorable conditions. De Jussieu[14] reports that in irrigated groves in India, 16-year-old neem trees reached diameters over 40 cm, but trees grown in this way break more easily and are subject to wind damage. Under mediocre conditions, the average diameter of trees in a 44-year-old grove was 25.5 cm and average height was 10.5 m. In Africa, it is generally assessed that at 1 year in good soil, a grove reaches a height of about 1.5 m, at 2 years a height of 2 m, and during the fourth year trees reach a diameter of 7 to 8 cm and a height of about 4.5 m.

Fishwick[13] notes that on poor sites, there was evidence that neem stagnated after the first 5 to 6 years, and for this reason rotations were reduced to 7 years, which appeared to be more profitable. Furthermore, it appears that the critical time in the development of the trees in the plantation occurs when crowns begin to touch: the third or fourth year with a spacing of 8 ft (2.44 m) × 8 ft, and fifth and sixth year at a spacing of 15 ft (4.57 m) × 15 ft; thus on poor sites the wider spaced trees would cost less to plant, would be larger, and would have a greater economic value. (Comparing two plantations on poor sites, the value of the trees at the wider spacing was four times greater.)

Because weeds compete for moisture and nutrients, early growth of seedlings is much retarded, and regular weeding and cultivation stimulates neem growth and vigor. Research carried out in Dehra Dun, India, has shown that seedlings that were weeded reached a height of 0.6 to 1.4 m (2 to 4.5 ft) by the end of the second season, but only 0.5 to 1 m (1.66 to 3.25 ft) if not weeded. Later, the seedlings that were not weeded were killed by weed suppression and frost.[1]

V. YIELD

According to Michel-Kim and Brandt,[18] the yield of neem varies between 10 and 100 tons of biomass (dried material) ha/year, depending upon rainfall, site conditions, and spacing, and 40 tons [12.5 m³ solid wood/ha/year (based on 1 m³ = 800 kg)] can be achieved easily under the proper conditions. About 50% of the biomass is contained in the leaves, about 25% in the fruit, and 25% in the wood.

Gravsholt et al.[9] reports that the first rotation yield in Ghana was 30 to 38 cords of fuelwood per hectare (approximately 13.5 to 17 m³/ha/year solid wood); and in Samaru in northern Nigeria, 7.5 to 67 cords (approximately 2.4 to 21 m³/ha/year). Commonly in West Africa, plantations are cropped on an 8-year rotation, with a spacing between trees of 8 × 8 ft (2.4 m × 2.4 m).

Fishwick[13] reports that on more suitable sites in Bornu Province in Nigeria, the yield was 15 to 27 cords/ha/year (13.5 to 24.3 m³ solid wood) on a 7-year rotation. The plantation spacing was 8 ft (2.43 m) × 8 ft, and site conditions included soils comprising drift sands with a pH ranging from 5.0 to 7.5 and a mean annual rainfall varying from 380 to 762 mm (15 to 30 in.). An average annual yield for all plantations was 7 to 15 cords/ha (6.3 to 13.5 m³). In Nigeria, neem poles are in great demand for house construction and fence posts because they are semiresistant to termites; poles thus realize a greater price than fuelwood.

McComb[19] reports plantation yields in Samaru, Zaria, Nigeria, of 300 to 2250 ft³/acre for the first 8-year rotation crop (2.55 to 19.67 m³/ha/year solid wood); and 350 to 2250 ft³ (3 to 19.67 m³/ha/year) for the 8-year coppice crops on the same plantations. (The lower yields were for class IV sites, while the higher ones were for class I sites, and the spacing between trees was 2.4 × 2.4 m.) Interestingly, the first 8-year rotation crop on class I plantation sites showed an average incremental growth rate of 350 ft³/year (after the second year), while the coppice crop averaged only 250 ft³/year. There were no significant growth rate nor yield differences on the class IV sites, and on the class I sites the volume of the coppice growth by the third year was almost equal to the growth by the fifth year of the first 8-year rotation crop.

Radwanski[21] gives the yields of the Majiya plantation near Sokoto, Nigeria, as 520 ft³/acre of fuelwood after 7 years (5.2 m³/ha/year) and after coppicing, 820 ft³ solid volume of fuelwood and 290 ft³ solid volume of timber per acre (total yield of 8.6 m³/ha/year). The plantation was planted in 1945 with "open-root" neem seedlings (assumed to mean bare-root), spaced 6 × 6 ft (approximately 1.8 × 1.8 m). The average annual rainfall in this region was 31 in. (787 mm) with a maximum of 47 in. (1194 mm) to a minimum of 20 in. (508 mm).

Welle[17] reports that trees grown in plantations in Haiti at 4 years of age averaged a yield of 1 pole plus and 0.09 m³ of fuelwood, but the volume of the pole was not given. Radwanski[22] gives a volume of 1.05 m³ for neem poles in one plantation in Nigeria, while McComb[19] reports a pole volume ranging from 37 to 85% of the total wood produced.

VI. COPPICE GROWTH

Early growth is faster from coppice than from seedlings, reaching a height of 8.5 m (28 ft) in 3 years after cutting. However, in Nigeria, the height of 8-year-old coppice was reported to have only equaled that of 8-year-old trees started from seedlings. In plantations there, trees reached a height of 7 m (23 ft) in 3 years and 12 m (40 ft) in 8 years. Approximately 66% of height is achieved in the first 3 years after planting.[23] Fishwick[13] reports that plantations in Nigeria were cut to a stump height of 8 cm (3 in.), and coppice from such low stumps is less likely to suffer from wind damage than from higher stumps. Higher stumps also have a higher incidence of dying off. Grose[10] observed that by cutting at a stump height of 1 to 2 m, the number of poles produced by coppice will increase. Radwanski[7] recommends a coppice management system every 3 years for maximum biomass yield based on the observation that 66% of the growth of the tree occurs in the first 3 years. However, this merits further research, since much faster growth of coppice occurs once the root system is established, and a much shorter rotation may give a maximum yield.

VII. INTERCROPPING

Radwanski and Wickens[8] report, "Neem cannot be grown among agricultural crops since it will not tolerate the presence of any other species in its immediate vicinity, and if not controlled, may become aggressive by invading neighboring crops." Troup[1] reported that seedlings that were not weeded were suppressed and eventually killed because of weeds (and frost). But Makay[24] reported that even though neem seedlings had been hidden from August and September under a 10-ft (3-m) high crop of millet, neem did surprisingly well. The seedlings were still healthy, did not appear to have been retarded after the millet was harvested in October, and a healthy stand of neem was left behind. Others state that neem can be planted in combination with fruit cultures and crops for feeding cattle (e.g., *Pennisetum pedicellatum,* as suggested by Misra).[25] Also, recommendations have been made for combinations of neem with sesame, cotton, and hemp;[26] with peanuts, beans, sorghum;[7,27,28] with *Acacia arabica* (synonym *A. nilotica*) and cotton;[1] and with *Khaya senegalensis.*[29]

In Nigeria, a form of taungya was used for neem plantation establishment and farmers cultivated groundnuts, beans, and millet between the trees, but the forest department planted the trees. These plantations were superior in survival and quality to plantations established by other means, and the cost was much lower. In areas where neem plantations were cleared, groundnuts were cultivated and yielded three times the average for other fields.[13]

It is known that the compounds found in neem are not only effective against insect pests, and beneficial in the efficient use of nitrogen, but they also affect some fungi and bacteria. Thus, the tree may have a significant influence on the balance within the microfauna, fungi, and bacteria communities. Because plants depend upon a certain microfauna and a special complex of bacteria and fungi, it is possible that where neem changes the composition significantly, problems may arise. The effect of neem may be both positive and negative, and research is needed to prove or disprove these factors.[18]

Perhaps neem is allelopathic to some crops. There are many conflicting statements as to its compatibility for intercropping with food crops; some agree that it has poor agroforestry potential because of its interference with other crops or vice versa. There is no clear explanation made as to the intolerances, either of neem to other crops or vice versa. Or the reason may be that since neem fruits produce a systemic, somewhat repugnant chemical, food crops may take up this chemical from fruits falling on the ground once the tree begins bearing fruit (usually at 5 years). Food crops might then have a bitter taste; hence the reference that neem is not a good species for agroforestry. Research is surely needed to prove or disprove the incompatability of neem as an intercrop and agroforestry species.

The answer may be to plant neem in mixed forests in combination with pasture. Michel-Kim and Brandt[18] suggest that up to 20% of the area could be planted to neem, and village plantings could constitute up to 15%. Neem would also be a good species for use in a sequential agroforestry system as illustrated in the Radwanski and Wickens paper.[8] Another excellent use of neem in agroforestry systems is for windbreaks. In the Majjia Valley in Niger, over 500 km of windbreaks (a form of agroforestry) comprising double rows of neem trees have been planted to protect millet crops and to supply wood to local villagers. This has resulted in a 20% grain yield increase for local farmers and the windbreaks are lopped and provide needed fuel and construction wood to villagers.[30,31]

Perhaps one of the best agroforestry potentials of neem is growing it for its various useful products where it is not intercropped with food crops, and the products (leaves, neem cake, etc.) are processed or used in a cut-and-carry system and applied to food crops as a fertilizer and pesticide.

VIII. NEEM SEEDS AND PROPAGATION

Neem begins to bear fruit in 3 to 5 years.[32] The period of collection of neem fruits naturally varies from place to place, depending upon the regional climatic conditions. In India, collection may begin as early as May and extend through September; however, there seems to be two distinct fruiting periods, May — June and August — September. The fruits are collected from the trees when fully ripe or are gathered from beneath trees.

A. YIELD

Neem produces fruits in 3 to 5 years and becomes fully productive in 10. Ketkar[2] gives a yield of about 50 kg (110 lb)/tree/year; Ahmed and Grainge[3] 30 to 50 kg (66.6 to 110 lb); and Radwanski[33] lists 11.4 to 34 kg (25 to 75 lb), averaging about 20.5 kg (45 lb). About 2000 to 3000 fruits weigh 1 kg, the depulped fruit yields about 1800 seeds/kg,[14] and 9 to 10 dry seeds weigh 1 g.[34] The ratio of seed to pulp is approximately 1:2, and fruit pulp and kernels account for 47.5 and 10.1%, respectively, of fruit weight.[35] For reproduction seed preparation, the fruits are rubbed and washed to remove the flesh from around the seeds. After washing, the seeds are dried in the shade and preferably stored in dry, airtight containers.

B. VIABILITY

Usually, neem seeds remain viable for only a few weeks, about 1 to 2 months, and normally they are collected when thoroughly ripe and sown as soon as possible. But when mature seeds are depulped and adequately dried and cooled, they can be stored for longer periods. Reportedly, germination rates rapidly decline during storage. However, Brouard[36] cites work by Dr. Paul B. Tompsell at the Royal Botanical Gardens in Great Britain, who froze seeds after drawing the seed moisture content to below 8% and they remained viable up to 2 years. To obtain a high germination percentage, seeds must be collected when they are fresh, cleaned thoroughly, and handled carefully (to avoid cracking).

C. GERMINATION

Seeds germinate in about 2 weeks after sowing. Fresh neem seed germinates quite readily and scarification is generally not needed. Research at the Royal Botanic Gardens in Great Britain indicates that germination is improved if the inner shell is removed to expose the embryo before planting,[8] and Smith[37] reports the same. However, Singh[34] recommends that the seeds be cut across with a sharp blade and the cotyledons examined: if the cotyledons are green the seeds are sound, but if they have turned yellow or brown, they will not germinate.

In more efficient containerized nurseries, it is desirable to produce a sprouted seed in every container because less space is required and it reduces per-seedling labor costs. Experiments were conducted by Operation Double Harvest in Haiti in an attempt to pregerminate neem seeds for transfer to container; however, the research showed that 47% of pregerminated seed had a major taproot deformity (crooked or looped), while the roots of only 7% of dry sown seed were deformed.[38]

Fagoonee[39] found that germination is best when seeds that have fallen from trees during the preceding week are soaked in water for about 3 d, depulped and cleaned, then sown directly into the nursery in damp soil. The soaking breaks the dormancy by neutralizing the germination-inhibiting chemicals found in the shell of the seed.

However, Ezumah[40] concludes that neem seeds do not require a period of after-ripening. Seed origin, year, and time of producton have no significant effect on germination and longevity. Sun-drying does not adversely affect viability and appears preferable to air-drying to bring seed moisture content to 10% or less before storage. The method of seed cleaning also has no significant effect on germination and longevity; thus, decomposing the pulp before washing by keeping the fruits in a heap is easier than peeling the pulp from fresh fruits. Cold storage adversely affects the viability of neem seed; seeds stored at room temperature (20 to 28°C) retained some viability for 16 weeks, while viability lasted for only 12 weeks at cold storage (6 to 7°C).

IX. PROPAGATION TECHNIQUES

Most commonly, neem seedlings are propagated in the nursery and transplanted to the field, although direct sowing has been successful in some areas if rainfall is adequate.[1,17]

Although neem needs light, young seedlings can suffer from strong solar radiation; thus a light shade is desirable during the first season of growth. Sudden exposure of seedlings without first hardening off will result in a high rate of mortality. Seedlings naturally regenerated under old stands often die when the trees are cut and the canopy is opened.[13]

A. NURSERY CARE
In the nursery, seedlings are either grown in containers (plastic bags or root-trainers) or in seedbeds. Germination starts in about 8 d and continues to about 3 weeks.

1. Bags and Other Containers
Seeds are sown directly into the container filled with potting soil and are ready for transplanting in 12 weeks.[41] In Haiti, seedlings are grown in containerized nurseries and transplanted in 3 to 5 months, depending upon the rainfall pattern.[17]

2. Seedbeds
If the nursery is irrigated, seeds can be harvested early, in April, and planted in the nursery. The seed should be lightly covered with earth and sparingly watered, the soil kept loose to prevent caking. Singh[34] recommends a spacing of 5 cm in-row and 20 cm inter-row, planting the seed at a depth of no less than 1 cm to minimize rodent damage. Fishwick[13] states that seeds should be sown thickly in lines 30 cm apart, selectively thinned when the seedlings are about 8 cm high (3 in.) to a spacing of 8

cm, and selectively thinned again in 4 to 5 months with only the best stock remaining at a spacing of about 23 cm (9 in.).

3. Seedlings

Recommendations when seedlings are ready and what the height is at the time of transplant vary. If seeds are planted early in April, seedlings should be ready for transplanting by July (4 months), reaching a height of 15 to 20 cm. De Jussieu[14] and Troup[1] report that neem fruits from April to July, and seeds are planted in seedbeds in a partially shaded nursery as soon as possible after harvest, around the middle of July when the rains begin, and are ready for transplant by the next rainy season, after reaching a height of about 0.80 to 1 m. Radwanski[7] and TAREC[41] indicate that seedlings are ready for transplant in about 12 weeks when they are 7.5 to 10 cm high (3 to 4 in.), with a taproot approximately 15 cm long (6 in.). Mitra[42] recommends transplanting when seedlings reach 7.5 to 10 cm. (3 to 4 in.) high. According to Singh,[34] seedlings transplanted younger than 1 year have a poor survival rate; this does not hold true according to experience in Haiti, where the survival rate of 3 to 5-month container-grown neem transplants is 85%.[17]

4. Root Balls

Seedlings that have not been planted during the first year are kept until the following season and sometimes until they are 2 years old. Year-old seedlings are carefully uprooted from the nursery leaving a ball of soil around the roots, and transplanted as soon as possible. Planting is done in July — August in pits dug in April — May, which allows weathering of the soil, and if rain immediately follows, good survival rate is ensured. In dry areas, about 90% of the leaves are removed, reducing evapotranspiration, and decreasing transplant shock. The areas in the vincinity of the trees must be kept weed-free.[34]

5. Stumps

In India, stumps are usually prepared from 2-year-old seedlings, although in irrigated nurseries, year-old seedlings may attain the desired size for stump preparation. The seedlings should be uprooted with care to avoid splitting or breaking the taproot. Stumps are prepared with 22-cm roots and 5-cm shoot portions and wrapped (bare-rooted) in moist gunney sacks and kept in the shade until transplanted. Just before planting in 30 cm³ pits, desiccated root and shoot tips are pruned.[34] In Nigeria, Fishwick[13] found that seedlings cut to a stump height of 0.3 m (1 ft) were better than taller ones because they were easier to handle and provided sufficient buds for sprouting, and were not damaged by wind, which whipped taller ones, loosened the soil, and exposed the roots. Roots were trimmed to 30 cm (12 in.), with most of the new growth produced by the callus that forms at the point where the roots are trimmed. Survival was increased when the seedlings were cut in the seed bed rather than at the time of lifting because shock was reduced by not cutting, lifting, and transplanting at the same time, and a callus tissue would form on the shoot wound before lifting. In Africa, Gorse[10] cut canes to a height of 1.5 m and removed all leaves before transplant, and in this way the bark is already hardened and not as susceptible to damage by animals.

6. Disadvantages

There are distinct disadvantages to planting overage seedlings. First, it is very costly to keep and care for seedlings for 1 to 2 years. The added weight of seedlings with a soil root ball drastically increases transportation costs and reduces the number of seedlings that can be delivered to the planting site, especially if they are hand carried. Another problem: when seedlings are transplanted bare-rooted, they suffer severe transplant shock, roots dry out, root hairs and beneficial mycorrhizae die. Survival rate is decreased, and growth rate is retarded.

B. TISSUE CULTURE

The use of tissue culture and cuttings to produce neem for reforestation has been deemed unrealistic in most cases because of the high cost of a production facility, and the relative ease of producing seedlings.[43] Also, it is questionable whether tissue-cultured plantlets and cuttings will develop full taproots. However, it could be economically justified to screen and reproduce plants of unique germplasm (such as a tree with an unusually high azadirachtin content) in this manner for seed orchard establishment in more moist areas, which would not be affected by limited taproot development.

Neem has been successfully tissue-cultured from leaflet callus tissue and from stem tissue by growing it in modified Murashige and Skoog media, producing roots in 40% of the cultures and developing into complete plantlets in a supplemented medium.[44-46] In further tissue culture research, Schultz[47] was not able to achieve root initiation as reported by Sanyal et al.,[44] which supported research by Rangaswamy and Promila[48] describing the differentiation of "growth centres" in *Azadirachta* callus and the eventual shoot bud formation with rarely any rooting. Again, it is questionable whether a normal length taproot will develop from tissue-cultured plantlets, which would exclude their planting in water-stressed areas.

C. CUTTINGS

Neem can also be propagated by cuttings, which require a production period of 6 months to 1 year,[41] but again the development of a normal taproot would be doubtful. Air-layered branches treated with IBA or NAA in lanolin paste at 0.1% develop roots satisfactorily.[49] In Nigeria, Fishwick[13] records that in Maiduguri a number of neem shoot cuttings were treated with a rooting hormone (Seradex B) and then placed in pots. A number took, but did not survive after being transplanted. In Sokoto, the work was repeated, but cuttings were covered with polythene bags, and a number survived after being transplanted. It was found that a significantly higher proportion of the cuttings took root when they were taken and prepared at the start of the rainy season.

D. DIRECT SEEDING

In the literature, the term "direct seeding" (sowing) is used in two ways — direct seeding in the nursery and direct seeding in the field. In India and Nigeria, direct sowing is reported to have proven more successful for reforestation than transplanting; however, this is when areas to be sown are well prepared (similar to land preparation for sowing food crops). Direct seeding on hard and inhospitable soils has not been very successful, but establishment and growth can be greatly enhanced when seeds are sown in bore holes that have been dug in these hard soils and filled with a fertile potting mix. Research has shown that hole size on difficult sites can significantly enhance establishment and growth of transplanted seedings. Although drilling bore holes with a post-hole auger and filling them with a potting mix may to some seem expensive, the cost is comparable to nursery raising and transplanting seedlings.

When direct-seeded, neem establishes an extensive root system before aerial growth becomes rapid.[4] Site preparation and transplanting are the two biggest costs of reforestation, and direct seeding can markedly reduce transplanting costs. De Jussieu[14] reports that aerial seeding has been tried but yielded poor results, since the chances of survival for the seeds is much lower than direct seeding on prepared land. Gorse[10] recommends the use of a groundnut planter and weeder for direct seeding and weeding, and covering the seed beds with a 10 to 15-cm mulch of groundnut hulls.

Welle[17] reports that direct seeding has proven to be a viable method of establishment in Haiti. Seedlings raised in nursery beds are ready for tranplanting during the first rains when they are 7.5 to 10 cm (3 to 4 in.) high. Sometimes seedlings are retained in nursery beds and transplanted during the second rainy season.[11]

Methods of direct seeding vary.[7,11,34] In India:

1. The soil is worked to a depth of about 15 cm and the seeds are sown at a depth of 1.5 cm. Sowing is done either in patches or lines; in the former they are spaced about 3 × 3 m, and spacing between lines is about 3 m, using 3 to 4 kg of seed per hectare. Weeding is necessary and seedlings are thinned at the end of the first season, leaving two seedlings per patch or two seedlings per meter length of the line.

2. Seed was sown at high and dry sites that had been plowed twice. No watering was done, but the seedlings were kept free of weeds. In less than 3 years, the plants were 7 to 8 ft high, the growth being equal to or better than that of transplants that had been carefully watered and tended. Trees in similar plantings at another site measured up to 12 ft in 3 years.

3. Direct seeding into mounds of earth 12 × 4 × 1.5 ft on sites receiving 24 in. (600 mm) of rain annually produced plants that reached a maximum height of 4.5 ft in 1 year.

4. Neem was sown in combination with the cultivation of sesame, cotton, and the lesser hemp in an area with an annual rainfall of less than 500 mm (20 in.). The sown lines were 0.3 m (1 ft) apart, three lines of field crops to one line of trees, so that the latter were 1.2 m (4 ft) apart. Sowing of both field crops and trees was done after site preparation by plowing and harrowing. The lines of trees were weeded twice during the first rains. The trees reached a height of 5 m (16 ft) and girth of 43 cm (17 in.) in 3 years.

5. Direct sowing into plowed furrows in black cotton-soil produced trees with a maximum height of 1.5 m (5 ft) after 1 year.

6. Success has been achieved by dibbling neem seed under Euphorbia bushes.

In Nigeria, Fishwick[13] reports successful direct seeding at the bases of the native cover, with a survival rate of about 40 trees/ha. It was observed that although the rate of growth is generally slower than on cultivated sites, it had merits of simplicity and cheapness to enrich degraded forest area.

In northern Nigeria, neem interplanted on farms among groundnuts, beans, and millet showed markedly superior growth. When the crop was harvested, a healthy stand of neem seedlings was left behind.[4]

E. SPACING

Neem seems to be very nutrient-demanding, and has been known to send out lateral surface roots reaching over 18 m.[50] In Nigeria, a spacing of 1.8 × 1.8 m is recommended,[14] and Gorse[10] recommends a much wider spacing on poorer sites.

F. ROOT FORMATION

Roots that have been pruned (as suggested in the stump planting method) may not regenerate into a long, normal taproot (rather they develop several shorter pseudo taproots, which are not as long as a normal one). Fishwick[13] found, by digging, that a 2-year-old tree planted from a stump possessed seven taproots, each with a 1.27-cm (1/2-in.) diameter at a depth of 12 ft (full depth of penetration was not recorded). The author does not know if neem, under normal circumstances from a seed with an undamaged root system, would develop only a single, deep taproot. Without a normal taproot, the plant would be limited in its ability to penetrate into low water tables and reach nutrients in lower soil horizons. During drought, when water tables drop considerably, the trees may die. Similar difficulties would be encountered when trees are established from cuttings, for they do not generate normal taproots. This might explain the die-off of the neem plantations (established by stump cuttings and from seedlings with the taproot pruned) in Maiduguri, Nigeria, reported by Fishwick[13] and elsewhere by others.[50]

G. SURVIVAL

The rate of seedling survival is influenced by a number of factors besides age of seedlings and methods of transplanting. Genetics play an important role, and seeds should be gathered from plus trees of the desired phenotype and ecotype. Also the size of the seed influences survival and early plant growth; larger seeds produce a much stronger seedling (maternal influence). Seedlings with a well developed root system (such as those promoted by "root trainers," and through fertilization, and those which have been inoculated with mycorrizae) can withstand drought much better and have the capacity for a more immediate and larger uptake of moisture and minerals, thus will have a higher survival rate.

H. NATURAL REPRODUCTION

Under natural conditions, the fruit (seed) ordinarily drops to the ground during the rainy season, and germination takes place in 1 to 2 weeks. Neem reproduces naturally with tolerable freedom, especially around trees growing in moist, sandy soils. Naturally regenerated seedlings have been used for reforestation, but they do not compare in vigor with good nursery stock. Their root systems are poorly developed, they are very sensitive to sunlight, and they lack buds.[13] In Haiti, a number of volunteer seedlings were dug up from under a tree; 22% were highly deformed, and only 39% could be rated as having well shaped taproots.[38] Neem establishes well under bushes and scrub,[1] though initial growth is usually slower.[4] Bats and birds are reported to spread neem by eating the fruit and depositing the seed elsewhere, and spontaneous individual trees and stands of neem trees are reported to have been established in this manner in several countries in Africa, India, and Haiti. Some feel that bats spread more seeds than birds, and larger numbers of volunteer seedlings can be found under trees where they roost.[10,14,51]

X. GERMPLASM

There is great variation in the germplasm of neem in terms of azadirachtin content, seed oil content, seed yield, form (clean, straight bole, branchy, etc.), fast-growth, tolerances to different environments, and resistance to diseases and pests. If neem is to be grown for pesticide production, it is necessary to develop trees that produce high yields of fruit with maximum azadirachtin content. Therefore, there is a great need to exploit the germplasm resources of neem in the area of its origin as well as in exotic regions, and to conduct research on its performance. There is also a need to broaden out the germplasm in exotic areas to avoid disease and pest infestations where it is widely planted. Since neem is native to the Burma-India area, all germplasm originates there.

There is a great variance in the azadirachtin content of the seed, which may depend upon a number of variables, and it is difficult to determine which one(s). The differences in azadirachtin content may be determined genetically and may be environmentally triggered or influenced, perhaps by water stress, high humidity, high or low levels of rainfall, or soil nutrient content. Azadirachtin content may also vary greatly depending upon the age of the tree, when the seed is picked, how it is dried, stored, shipped, exposed to light, heat or cold, etc. Some think azadirachtin content of seed from trees 5 or more years of age is higher than that from younger trees. The only way this can be determined is by scientifically noting the environment where the trees are growing (altitude, soil type, latitude, rainfall, and pattern — e.g., was the tree water-stressed before fruiting?), determining the age of the tree, collecting all seed at the same age or stage after fruiting (green, ripe but still on the tree, fallen to the ground), and cleaning, drying, storing, handling, and analyzing the seed in the same manner.

Jacobson[6] analyzed seed from some varieties from India that contained only 1 to 3% of the chemical, while seed from Africa tested at 5 to 6%, and some as high as 9%. Also, collections of neem seeds (reared from Togo-bred seeds) in June and September have shown that the seeds from older trees contain larger quantities of pesticidal compounds, especially azadirachtin.[51] Ermel et al.[52] analyzed seed from 66 sources. The highest contents were measured in samples from Nicaragua and Indonesia, showing 4.8 and 4.85% azadirachtin, respectively. High contents were also measured in kernels from Togo, India, Burma, and Mauritius (3.3 to 3.9%), wherease samples from Sudan and Niger gave lower yields (1.9% and 1.5%). The incubation of kernels under increasing temperatures and high air moisture resulted in time-dependent decrease of the azadirachtin contents. The exposure of extracts to sunlight and ultraviolet radiation also decreased the azadirachtin contents remarkably. Singh[53] notes that an attempt to correlate the antifeedant efficacy with extract yield and oil content and these three factors with environment often failed. However, it was observed that seeds from neem trees growing in dry areas near the desert possessed much higher biological activity than those from trees growing in coastal areas.

Reports on the amount of oil in the seed vary from as low as 17 to over 59%. According to Radwanski,[33] the kernels constitute about 45% of the fruit and yield about 45 to 49% oil (25% of the whole fruit); proper handling of the seed and more refined methods of extraction yield higher levels of oil. The yield of oil from the Ceylonese seed kernels is reported to be the highest at 59.25%. Larson[54] states there is no evidence at this time that the azadirachtin level rises linearly with the percentage of the oil since the azadirachtin does not tend to follow the oil in the extraction process using ethanol. Mitra[42] and Ketkar[2] estimate neem seed to contain about 20% oil, which contains about 2% active ingredients with manufacturing potential for producing pharmaceutical and insect-repellent preparations. Singh[53] notes that kernels from trees growing in humid areas or areas with more rainfall yielded more oil than those from other places. Again, the amount of oil in the kernel is probably inherited and may be environmentally influenced.

It is also necessary to develop fast-growing neem germplasm for two other purposes: (1) maximum clean bole height and (2) maximum branching at a low height that produce straight coppiced poles, in order to remit maximum income to farmers in developing countries.

XI. TOXICITY, PESTS, DISEASES, AND LIMITATIONS

Neem is said to have few pests and its naturally occurring pesticide is nontoxic to man and animals. However, there is evidence that this may not be so.

A. TOXICITY

Radwanski and Wickins[8] state that one of the most significant attributes of neem oil as an insecticide is that it is effective and reportedly nontoxic to man or animals, and nonpolluting to the environment. However, a syndrome similar to Reye's syndrome has appeared in children given large doses of neem oil;[55,56] although infrequent, this manifestation has been fatal. Because Reye's syndrome and the neem oil-induced Reye's-like syndrome are poorly understood, neem extracts should be tested more thoroughly before being used for medicinal purposes. Also, neem extracts have been found to be toxic to guinea pigs and rabbits,[57] the insectivorous fish *Gambusia* spp., and tadpoles died at 0.04% concentration of neem extracts.[58] Welle[17] observed that neem seeds falling into fish ponds in Haiti proved fatal to Talapia fry. This indicates the need for systematic follow-up toxicological studies.

Pereira and Wohlgemuth[59] report that neem seeds carry *Aspergillus flavus*, which in certain conditions produces aflatoxins; this is a particular concern in waste piles at preparation sites.

B. ALLELOPATHY

There is some concern that neem compounds may be allelopathic, but no research evidence supports this contention (also see Section VII).

C. PESTS

Roberts[60] lists 14 insect species and 1 parasitic plant as recorded pests of neem in Nigeria, although few are serious, and most plantations of neem are reported to be insect-free, evidently due to the repellant compounds of the tree. Insects will eat the radicle of germinating seeds if it is not covered well with soil. Occasional insect infestations by species of *Microtermes* and plant parasitism by *Lorantium* have been observed in Nigeria, but the attacked trees almost invariably recover, though their rate of growth and branch development may be considerably retarded.[8] CAB-IIBC[61] reports that *Aonidiella orientallis* has become a pest of neem in some parts of Africa (e.g., the Lake Chad Basin: Niger, Chad, Nigeria, and Cameroon).

In India, the larvae of *Enarmonia koenigana* Fabr. feed on rolled leaves and bore into tender shoots, and the larvae of *Cleora cornaria* Meyrick and *Odites atmopa* defoliate the leaves.[62] Warthen[63] lists nine other insect pests that attack neem: *Calepiterimerus azadirachta, Araecerus fasiculatus, Cryptocephalus ovulus, Holotrichia consanguinea, H. insularis, H. serrata, Pulvinaria maxima, Laspeyresia aurantianna,* and *Orthacris simulans.*

Termites have been known to damage trees, attacking them at the level of the collet, sometimes extending to the trunk and to the cyme, but generally not killing them.[14] Fishwick[13] reports that termites attack weak, sickly trees but at times they also attack and kill young, vigorous trees. Neem coppice shoots have also been attacked and killed, but the roots of the stump were not damaged and produced fresh coppice shoots.

Gosh[64] reports that the neem scale, *Palvinaria maxima,* is a serious pest in central and south India. It feeds on sap, covers the tender shoots and stem in numbers, and sometimes damages a young tree considerably. A tree in an advanced stage of infestation is recognized by the thick coating of white, mealy patches formed on the foliage, shoots, and bark. Another scale insect, *Aspidiotus orientalis,* thickly covers the shoots and stems of seedlings of about 0.6 to 2.5 cm in diameter, appearing on the new shoots and spreading to the leaves. In severe infestations, the growth is retarded, leaves are shed, the stems die back, and young trees may be killed. Also, *Aspidiotus pseudocer-iferus* feeds on the sap, and the nymphs of *Helopeltis antonii* feed on sap by puncturing the soft plant tissue, which blackens and dries. The wounds cause deformation of leaf and shoot, or the whole shoot may dry up and die back.

D. DISEASES

Recorded pathogens attacking neem are *Ganoderma lucidum,* causing root rot; *Corticium salmonicolor,* causing stem and twig blight; *Cercospora subsessilis,* causing leaf spots; *Oidium* sp., causing powdery mildew; and *Pseudomonas azadirachtae,* causing leaf spot and blight.[65,66] Sankaran et al.[67] reports additional diseases attacking seedlings in nurseries in India: *Rhizoctonia solani,* causing web blight; *Sclerotium rolfsii,* causing stem rot; *Colletotrichum capsici* and *C. subsessilis,* causing leaf spot; and *Fusarium solani,* causing wilt. Stem rot, leaf blight, web blight, and wilt reported for the first time caused up to 30% mortality of 2 to 3-month-old seedlings.

Singh and Chohan[68] observed a severe canker disease of twigs and shot-holes in leaves of neem and identified a fungus, *Phoma jolyana,* as the cause. De Jussieu[14] reports that cankerous lesions can also sometimes be found on the tree in fissures coming up from the collet along the stem. The wood becomes brown around the cankers and this coloration can penetrate to the heartwood. Gorse[10] observed that canker prevailed on the sunset side of the tree. This disease seems to correlate with a sudden absorption of water after a long drought, but through exploration of the genetic potential of neem, genetic resistance might be found.

E. PARASITES

Mistletoes that parasitize neem include *Dendrophthoe falcata* and *Tapinanthus* sp.[69]

F. WEEDS AND TILLAGE

Neem is intolerant of grass competition and needs thorough weeding, especially in dry areas, to obtain good growth.[4] Thorough weeding without watering is found to gain results almost, if not quite, equal to those attained by irrigation and weeding. The loosening of soil to prevent caking (in some soils), which promotes soil aeration and increases moisture percolation, is found to be most beneficial. Research indicates that on undisturbed soil, less than 25% of the rain falling on it is absorbed, but up to 60% when frequently hoed; mulching plus hoeing increased the absorption up to 90%.[13]

However, Fishwick[13] concludes that once weeds and grasses are established and root below the upper zone of competition, their presence does not seem to affect the tree growth, although it should be noted that perennial grass roots have been observed at depths of over 3 m (10 ft).

Experiments in Nigeria have shown that some tillage during and at the end of the wet season has a remarkable effect on the growth, health, and survival of neem in the first year when interplanted among groundnuts, beans, and millet.[24]

In compacted and eroded soils, survival rate can be influenced by the size of the hole in which the seedling is planted. A larger hole allows for increased water infiltration because of the uncompacted soil, and it also allows better development of the root system. In nutrient-poor soils, it should be possible to gain a higher rate of establishment by boring or augering a hole into the ground about 10 cm (approximately 4 in.) in diameter by 15 to 23 cm (approximately 6 to 9 in.) deep, replacing the displaced soil with a good nutrient-rich nursery mix, and seeding directly into the hole. This would greatly reduce reforestation costs, doing away with the expensive nursery operation.

G. OTHER PROBLEMS

Neem seedlings are killed by frost and fire, and large trees are frequently snapped off during high winds. However, trees seemingly killed by fire will coppice and regrow if cut soon after the burn.[10] Regeneration beneath stands of neem is sensitive to sudden exposure to direct sunlight, and clear felling in plantations normally results in the death of this regeneration if seedlings are under 8 cm (3 in.) in height.[13]

In some localities rats and porcupines girdle neem seedlings and trees, gnawing the bark from around the base and killing them. In areas with high rodent densities, they devour the fruits greedily and consume most of them after they fall to the ground. Goats and camels have been known to severely browse young plants and kill them.[34]

Because of the potential of pest outbreaks, neem should not be raised in pure stands, and it is recommended that neem should be mixed with leguminous trees like *Leucaena leucocephala, Albizzia lebbeck,* or *Acacia nilotica,* which reduce the risks of infestation by these pests. These trees would also complement the non-nitrogenous neem tree. Welle[17] reports that neem interplanted with *Leucaena leucocephala* in Haiti grew better than neem in monoculture.

REFERENCES

1. **Troup, R. S.,** *The Silvicultured Indian Trees,* Vol. 1, Clarendon Press, Oxford, 1921.
2. **Ketkar, C. M.,** Utilization of Neem (*Azadirachta indica* Juss.) and Its By-Products, Final Tech. Rep., Directorate of Non-Edible Oils and Soap Industry, Khadi and Village Industries Commission, Hyderabad, India, 1976; as cited by **Radwanski, S. A. and Wickens, G. E.,** *Econ. Bot.,* 35, 398, 1981.
3. **Ahmed, S. and Grainge, M.,** The use of indigenous plant resources in rural development: potential of the neem tree, *Int. J. Dev. Tech.,* 3, 123, 1985.
4. **Anon.,** *Firewood Crops: Shrubs and Tree Species for Energy Production,* National Academy of Sciences, Washington, D.C., 1980.
5. **Pliske, T. E.,** The establishment of neem plantations in the American tropics, in *Natural Pesticides from the Neem Tree and Other Tropical Plants,* Schmutterer, H. and Ascher, K. R. S., Eds., GTZ Press, Eschborn, West Germany, 1984, 521.
6. **Jacobson, M.,** Neem research and cultivation in the Western Hemisphere, in *Proc. 3rd Int. Neem Conf.,* Nairobi, Kenya, 1986, Schmutter, H. and Ascher, K. R. S., Eds., GTZ Press, Eschborn, West Germany, 1987.
7. **Radwanski, S. A.,** Neem tree. I. Commercial potential characteristics and distribution, *World Crops Livestock,* 29, 62, 1977.
8. **Radwanski, S. A. and Wickens, G. E.,** Vegetative fallows and potential value of the neem tree (*Azadirachta indica*) in the tropics, *Econ. Bot.,* 35, 398, 1981.
9. **Gravsholt, S. et al.,** Provisional Tables and Yields of Neem (*Azadirachta indica*) in Northern Nigeria, Research paper No. 1, Savanna Forest Research Station, Samaru, Zaria, Nigeria, 1967, 37.
10. **Gorse, J. E.,** personal communication, 1986.
11. **Anon.,** Neem (*Azadirachta indica*), Indian

Central Oilseeds Committee, NE-Oilseed Series No. 2., Hyderabad, undated.
12. **Jackson, J. K.,** Irrigated Plantations, FAO Forestry Paper No. 11 (Savanna Afforestation in Africa), UN/FAO, Rome, Italy, 1976.
13. **Fishwick, R. W.,** Neem (Azadirachta indica Ard. Juss.) Plantations in the Sudan Zone of Nigeria, Rep. prepared for the Chief Conservator of Forests, Northern Nigeria, undated.
14. **De Jussieu, A.,** *Azadirachta indica* and *Melia azedarach* Linne: silvicultural characteristics and planting methods, *Rev. Bois Forets Trop.,* No. 88, 23, 1963.
15. **Zech, W.,** Investigations on the occurrence of potassium and zinc deficiencies in plantations of *Gmelina arborea, Azadirachta indica* and *Anacardium occidentale* in semiarid areas of West Africa, in *Potash Review,* No. 1, International Potash Institute, Berne, Switzerland, 1984.
16. **Senior Resident of Sokoto,** Sokoto Survey, Report prepared for the Development of Sokoto Province, Nigeria, 1948; as cited by **Radwanski, S. A.,** *World Crops Livestock,* 29, 222, 1977.
17. **Welle, P.,** Operation Double Harvest — Haiti, personal communication, 1985.
18. **Michel-Kim, H. and Brandt, A.,** The cultivation of neem and processing it in a small village plant, in *Natural Pesticides from the Neem Tree (Azadirachta indica A. Juss),* Schmutterer, H., Ascher, K. R. S., and Rembold, H., Eds., GTZ Press, Eschborn, West Germany, 1981, 279.
19. **McComb, A. L.,** Provisional Tables for Growth and Yield of Neem (*Azadirachta indica*) in Northern Nigeria, Research paper No. 1, Savanna Forestry Research Station, Samaru, Zaria, Nigeria, 1967.
20. **Kemp, R. H.,** Trials of Exotic Tree Species in the Savanna Region of Nigeria, Research

paper No. 6, Savanna Forestry Research Station, Samaru, Zaria, Nigeria, 1970.

21. **Radwanski, S. A.,** Neem tree. III. Further uses and potential uses, *World Crops Livestock,* 29, 167, 1977.

22. **Radwanski, S. A.,** Neem tree. IV. A plantation in Nigeria, *World Crops Livestock,* 29, 222, 1977.

23. Department of Forest Research — Nigeria, Provisional Tables for Growth and Yield of Neem *(Azadirachta indica)* in Northern Nigeria, Research paper No. 1, Savanna Forestry Research Station, Samaru, Zaria, Nigeria, 1967.

24. **Mackay, J. H.,** Notes on the establishment of neem plantations in Bornu Province of Nigeria, *Farm and Forest,* 11, 9, 1952; as cited by **Radwanski, S. A.,** *World Crops Livestock,* 29, 62, 1977.

25. **Misra, B. R.,** Creation of fuel-cum-fodder reserves in the Plains of Chattisgarh (Madhya Pradesh), Proc. Farm Forestry Symp., New Delhi, 1960; as cited by **Michel-Kim, H. and Brandt, A.** in *Natural Pesticides from the Neem Tree (Azadirachta indica A. Juss),* Schmutterer, H., Ascher, K. R. S., and Rembold, H., Eds., GTZ Press, Eschborn, West Germany, 1981, 279.

26. **Howaldt, T. H.,** *Azadirachta indica* A. Juss — *Tamarindu indica* L., M. S. thesis, University of Hamburg, West Germany 1980; as cited by **Michel-Kim, H. and Brandt, A.,** in *Natural Pesticides from the Neem Tree (Azadirachta indica A. Juss),* Schmutterer, H., Ascher, K. R. S., and Rembold, H., Eds., GTZ Press, Eschborn, West Germany, 1981, 279.

27. **Radwanski, S. A.,** Improvement of red acid sands by neem tree *(Azadirachta indica)* in Sokoto, Northwestern State of Nigeria, *J. Appl. Ecol.,* 6, 507, 1969.

28. **Radwanski, S. A.,** Trees to improve soil, *Nature,* 225, 132, 1970.

29. **Giffard, P. L.,** L'arbres de'Sénégal, Dakar, 1979; as cited by **Michel-Kim, H. and Brandt, A.** in *Natural Pesticides from the Neem Tree (Azadirachta indica* A. Juss), Schmutterer, H., Ascher, K. R. S., and Rembold, H., Eds., GTZ Press, Eschborn, West Germany, 1981, 279.

30. **Long, S. et al.,** Influence of a Neem *(Azadirachta indica)* Windbreak Plantation on Millet Yields and Microclimate in Niger, West Africa, Rep. to CARE International, New York, 1986.

31. **Bognetteau-Verlinden, E.,** Study of Impact of Windbreaks in Majjia Valley, Niger, M. S. thesis, 1980.

32. **Jacobson, M.,** personal communication, 1980.

33. **Radwanski, S. A.,** Neem tree. II. Uses and potential uses, *World Crops Livestock,* 29, 111, 1977.

34. **Singh, R. V.,** *Fodder Trees of India,* Oxford & IBH Publishing, New Delhi, 1982.

35. **Ketkar, C. M.,** Better utilization of neem *(Azadirachta indica)* A. Juss) cake, in *Proc. Int. Cong. on Oil and Oilseeds,* New Delhi, India, 1979.

36. **Brouard, J.,** personal communication, 1987.

37. **Smith, R.,** Royal Botanical Gardens, Wakehurst, Great Britain; as cited by **Radwanski, S. A. and Wickens, G. E.,** *Econ. Bot.,* 35, 398, 1981.

38. **Larson, J.,** 1st Quarter 1985 Operation double Harvest Rep. to USAID/Haiti, Agency for International Development, Washington, D.C.

39. **Fagoonee, I.,** Germination tests with neem seeds, in *Natural Pesticides from the Neem Tree and Other Tropical Plants,* Schmutterer, H. and Ascher, K. R. S., Eds., GTZ Press, Eschborn, West Germany, 1984.

40. **Ezumah, B. S.,** Germination of storage of neem *(Azadirachta indica* A. Juss) seed, *Seed Sci. Technol.,* 14, 593, 1986.

41. Forestry of Promising Species for Production of Firewood in Central America, Tropical Agronomical Research and Education Center, Tech. Rep. No. 86, Department of Renewable Natural Resources, Turrialba, Costa Rica, 1986.

42. **Mitra, C. R.,** Neem, Indian Central Oil Seeds Committee, Hyderabad, India, 1963; as cited by **Radwanski, S. A. and Wickens, G. E.,** *Econ. Bot.,* 35, 398, 1981.

43. **Sommer, H. E. and Caldas, L. S.,** In vitro methods applied to forest trees, in *Plant Tissue Culture, Methods and Applications in Agriculture,* Academic Press, New York, 1981; as cited by **Pluymers, D. W.,** Deployment of the neem tree in the arid third world for fuel, pharmaceuticals, and insecticides, presented at the 13th Annu. Third World Conf., Chicago, 1987.

44. **Sanyal, M. et al.,** In vitro hormone induced chemical and histological differentiation in stem callus of neem *Azadirachta indica* A. Juss, *Indian J. Exp. Biol.,* 19, 1067, 1981.

45. **Jaiswal, V. S. and Narayan, P.,** Regeneration of plantlets from stem tissue of *Azadirachta indica* Juss, *Proc. Natl. Sum. on Applied Biotechnology of Medicinal Aromatic and Timber Yielding Plants,* Datta, P. C., Ed., 1984, 132.

46. **Narayan, P. and Jaiswal, V. S.,** Plantlet regeneration from leaflet callus of *Azadirachta indica* Juss, *J. Tree Sci.,* 4(2), 1986.

47. **Schultz, F. A.,** Tissue culture of *Azadirachta indica,* in *Natural Pesticides from the Neem Tree and Other Tropical Plants,* Schmutterer, H. and Ascher, K. R. S., Eds., GTZ Press, Eschborn, West Germany, 1984, 521.

48. **Rangaswamy, N. S. and Promila,** Morphogenesis of the adult embryo of *Azadirachta indica* A. Juss, *Z. Plant Physiol.,* 67, 377, 1972; as cited by **Schultz, F. A.,** in *Natural Pesticides from the Neem Tree and Other Tropical Plants,* Schmutterer, H. and Ascher, K. R. S., Eds., GTZ Press, Eschborn, West Germany, 1984, 521.

49. **Shanmugavelu, K. G.,** A note on the air-layering of eugenia (*Eugenia jambolana* Lam) and neem (*Azadirachta indica* A. Juss), *South Indian Hortic., Coimbator,* 15, 70, 1967; as cited by **Singh, R. V.,** *Fodder Trees of India,* Oxford & IBH Publishing, New Delhi, 1982.

50. **Labelle, R.,** International Council for Research in Agroforestry, presentation at the National Academy of Sciences, unpublished report, 1985.

51. **Brouard, S. and Brouard, J.,** personal communication, 1987.

52. **Ermel, K. et al.,** Azadirachtin content of neem kernels from all over the world and its dependence on temperature, air, moisture and light, Abstr. 3rd Int. Neem Conf., Nairobi, Kenya, July 1986, 25.

53. **Singh, R. P.,** Comparison of antifeedant efficacy and extract yields of neem (*Azadirachta indica* A. Juss) ecotypes and parts of a neem tree, Abstr. 3rd Int. Neem Conf., Nairobi, Kenya, July 1986, 26.

54. **Larson, R.,** The Development of Margosan-O™, a Shelf-Stable Pesticide Non-Toxic to Humans, Domestic Animals, Birds and Honeybees, Made From an Ethanolic Extract of Neem Seed, U. S. Patent 4,556,562, 1985, Abstr. 3rd Int. Neem Conf., Nairobi, Kenya, July 1986, 31.

55. **Sinniah, D. et al.,** Reye-like syndrome due to margosa oil poisoning: report of a case with postmortem findings, *Am. J. Gastroenterol.,* 77, 158, 1982.

56. **Sinniah, D. et al.,** Investigation of an animal model of a Reye-like syndrome caused by margosa oil, *Pediatric Res.,* 19, 1346, 1985.

57. **Sadre, N. L., Vibhavari, Y., Deshpande, U., Mendulkar, K. N., and Nandal, D. H.,** Male antifertility activity of *Azadirachta indica* in different species, in *Natural Pesticides from the Neem Tree and Other Tropical Plants,* Schmutterer, H. and Ascher, K. R. S., Eds., GTZ Press, Eschborn, West Germany, 1984, 473.

58. **Jotwani, M. G. and Srivastava, K. P.,** Neem, pesticide of the future. II. Protection against field pests, *Pesticides,* 15, 40, 1981.

59. **Pereira, J. and Wohlgemuth, R.,** Neem (*Azadirachta indica*) of West African origin, *J. Appl. Entomol.,* 94, 208, 1982.

60. **Roberts, H.,** A preliminary Check List of Pests and Diseases of Plantation Trees in Nigeria, Federal Dept. Forest Research, Ibadan, Nigeria, 1965; as cited by **Kemp, R. H.,** in Research paper No. 6, Savanna Forestry Research Station, Samaru, Zaria, Nigeria, 1970.

61. CAB International Institute of Biological Control (CAB-IIBC), Prospects for Biological Control of the Oriental Yellow Scale, *Aonidiella orientalis* (Newstead) as a Pest of Neem in Africa, Ascot, Berks, England, 1987.

62. **Bhasin, H. et al.,** A list of insect pests of forest plants in India and the adjacent countries. III. List of insect pests of plant genera A, *Indian For. Bull.,* [N.S.], Entomology, Manager of Publications, Delhi; as cited by **Singh, R. V.,** *Fodder Trees of India,* Oxford & IBH Publishing, New Delhi, 1982.

63. **Warthen, J. D., Jr.,** *Azadirachta indica:* a source of insect feeding inhibitors and growth regulators, U. S. Department of Agriculture, Reviews and Manuals, ARM-NE4, 1979.

64. **Gosh, R. C.,** *Azadirachta indica* A. Juss, 1984, unpublished paper.

65. **Bakshi, B. K.,** *Forest Pathology, Principles and Practice in Forestry,* Controller of Publications, Delhi, 1976.

66. **Desai, S. G. et al.,** A new bacterial leaf spot and blight of *Azadirachta indica* A. Juss, *Indian Phytopathol.,* 19, 322, 1966; as cited by **Singh, R. V.,** *Fodder Trees of India,* Oxford & IBH Publishing, New Delhi, 1982.

67. **Sankaran, K. U. et al.,** Seedling diseases of *Azadirachta indica* in Kerala, India, *Eur. For. Pathol.,* 16 (5/6), 324, 1986.

68. **Singh, I. and Chohan, J. S.,** *Phoma jolyana,* a new pathogen on neem (*Azadirachta indica*), *Indian Forester,* 1058, 1984.

69. **Browne, F. G.,** *Pests and Diseases of Forest Plantation Trees,* Clarendon Press, Oxford, 1968; as cited by **Singh, R. V.,** *Fodder Trees of India,* Oxford & IBH Publishing, New Delhi, 1982.

2 The Chemistry of the Neem Tree

Philip S. Jones
Roche Products Ltd.
Welwyn Garden City, England

Steven V. Ley
Department of Chemistry
Imperial College of Science and Technology
South Kensington, London, England

E. David Morgan
Department of Chemistry
University of Keele
Staffordshire, England

Dinos Santafianos
Department of Chemistry
Imperial College of Science and Technology
South Kensington, London, England

TABLE OF CONTENTS

I. INTRODUCTION

The unusual biological properties of the neem tree (*Azadirachta indica* A. Juss family Meliaceae), well described elsewhere in this volume, have been known in the Indian subcontinent for many centuries but only attracted the attention of scientists in the past 50 years or less. However, the value of the tree as one resistant to drought and which forms a beautiful shade canopy was recognized in British Colonial times, so the tree was spread throughout the tropical regions, particularly to East and West Africa and the Sudan and is now found growing from Guiana to Australia. Material from the tree was therefore widely available for investigation.

As is usual with such botanical material, the bark, heartwood, leaves, fruit, and seed have been examined to some extent by chemists (see Table 1), but it is the renewable parts, the leaves and seeds, that have received most attention. However, even today, our knowledge, as this review will show, is still only fragmentary. The impetus towards the investigation of triterpenoids of neem was given by the first isolation of a substance with insect feeding-deterrent properties.[1] This substance, azadirachtin (Structure 1)* is a highly oxidized triterpenoid. Today, we have an increasing body of knowledge of these fascinating and structurally complex substances but the study of other types of compounds from the neem tree has hardly begun.

The seed oil, extracted by steam or solvents from the crushed seeds, is a commercial product in India. It is a thick dark brown semisolid with a bitter taste and a smell that combines the odors of peanuts and garlic. It consists chiefly of triglycerides and large amounts of triterpenoids. Because of its insect feeding- and growth-disruptant properties, the crude seed oil has become the chief material of study, followed by the leaves. Some early work has been done on the triglyceride fraction of the oil, but this is very qualitative. Gas chromatographic analysis showed the glycerides contain chiefly oleic acid, followed by stearic and palmitic acids.[2] Some preliminary work has been done on phenolic compounds extractable from the bark and two substances, sugiol (Structure 2a) and nimbiol (Structure 2b) have been isolated and identified[3] these appear to be derived from terpenoid sources. The garlic odor of the seeds and oil and the presence of sulfur (as an impurity) in solids obtained from the oil,[4] suggest the presence of compounds similar to diallyl disulfide and may in turn suggest that substances like ajoene,[5] a recently discovered substance with antithrombotic activity, may be present.

II. TRITERPENOIDS

All of the well-characterized compounds identified in neem belong to the class of triterpenoids. From diverse biosynthetic studies in plants, these are known to be derived from acetate, via mevalonate and squalene, which is then cyclized. From the stereochemical arrangement of the methyl groups at C-10, C-13, and C-14 and the side chain at C-17, the known triterpenoids of neem are all derived from the parent tetracyclic triterpenoid tirucallol (Structure 3) (20-epi-euphol or kanzuiol), with a tirucallanane structure ($5\alpha,13\alpha,14\beta,17\alpha,20S$-lanostane).[6] Tirucallol itself has not been isolated from any neem product but all the substances that have been characterized, and which form the bulk of this review, can be considered as successive rearrangement and oxidation products of it. These can be divided into triterpenoids with the side chain intact, those in which carbon atoms 24 to 27 have been lost (tetranortriterpenoids), and further altered products (penta- and hexanortriterpenoids). Those in which the remaining carbon atoms

---

* Structures for this chapter appear at the end of the text.

Table 1
SUBSTANCES ISOLATED FROM VARIOUS PARTS OF THE NEEM TREE

Name	Structure	M.P.	Mol. Formula	Mol. Wt.	Source[a]	Isol.	Ref.
Azadirachtanin A	23	225	$C_{32}H_{40}O_{11}$	600	L	19	19
Azadirachtin	1	165	$C_{35}H_{44}O_{16}$	720	S	1	35,44
3-Desacetyl-3-cinnamoyl-	58	—	$C_{42}H_{48}O_{16}$	808	L	44	44
22,23-Dihydro-23β-methoxy	59	—	$C_{36}H_{48}O_{17}$	752	S	43	43
1-Tigloyl-3-acetyl-11-methoxy azadirachtinin	57	—	$C_{36}H_{46}O_{16}$	734	B	44	44
Azadirachtol	61	—	$C_{32}H_{46}O_6$	526	F	62	62
3-Tigloylazadirachtol	60	204	$C_{33}H_{42}O_{14}$	662	S	44	44
Azadiradione	8	—	$C_{28}H_{34}O_5$	450	S	11b	11b
7-Desacetyl-7-benzoyl-	9	—	$C_{33}H_{36}O_5$	512	S	15	15
7-Desacetyl-7-benzoylepoxy-	11	—	$C_{33}H_{36}O_6$	528	S	15	15
7-Desacetyl-7-hydroxy-	—	160	$C_{26}H_{32}O_4$	408	F	65	65
1β,2β-Diepoxy-	15	110	$C_{28}H_{34}O_7$	482	S	15	15
4α,6α-Dihydroxy-A-homo-	41	177	$C_{28}H_{36}O_6$	468	L	32	32
17-Epi-	13	—	$C_{28}H_{34}O_5$	450	S	12	12
Epoxy-	10	199	$C_{28}H_{34}O_6$	466	S	11b	11b
17β-Hydroxy-	12	—	$C_{28}H_{34}O_6$	466	S	12,57	12
1α-Methoxy-1,2-dihydro-	16	235	$C_{29}H_{38}O_7$	498	S	15	15
Azadirone	7	—	$C_{28}H_{36}O_4$	436	S	11b	11b
Gedunin	42	218	$C_{28}H_{34}O_7$	482	S	11b	—
7-Desacetyl-	44	—	$C_{26}H_{32}O_6$	440	S	11a	11a
7-Desacetyl-7-benzoyl-	43	278	$C_{33}H_{36}O_7$	544	S	15	15
Isoazadirolide	53	—	$C_{32}H_{42}O_{10}$	586	L	37	37
Isonimbocinolide	63	—	$C_{32}H_{42}O_9$	570	L	65	65
Isonimolicinolide	51	100	$C_{30}H_{38}O_8$	526	F	36	36
Margosinolide	64	130	$C_{27}H_{32}O_8$	484	T	68	68
Iso-	—	—	$C_{27}H_{32}O_8$	484	T	68	68
Meldenin	17	240	$C_{28}H_{38}O_5$	454	S	13	13
Meliantriol	5	176	$C_{30}H_{50}O_5$	490	S	7	7
Nimbandiol	38	121	$C_{26}H_{32}O_7$	456	S,L	31	31
6-Acetyl-	39	178	$C_{28}H_{34}O_8$	498	S	66	66
Nimbidinin	34	282	$C_{26}H_{34}O_6$	442	S	58,59	58,59
Nimbin	22	201	$C_{30}H_{36}O_9$	540	SBW	60	60
Desacetyl-	24	208	$C_{28}H_{34}O_8$	498	SBF	60	60
4-Epi-	—	—	$C_{30}H_{36}O_9$	540	S	69	69
Photooxidized-	47	184	$C_{30}H_{36}O_{10}$	556	S	13	13
Nimbinene	36	134	$C_{28}H_{34}O_7$	482	SLB	31	31
6-Desacetyl	37	141	$C_{26}H_{32}O_6$	440	SLB	31	31
Nimbiol	26	250	$C_{18}H_{24}O_2$	272	B	3	3
Nimbocinone	6	76	$C_{30}H_{46}O_4$	470	L	8	8
Nimbolide	35	245	$C_{27}H_{30}O_7$	466	L	30	30
Nimbolin A	31	180	$C_{39}H_{46}O_8$	642	W	29	29
Nimbolin B	32	243	$C_{39}H_{46}O_{10}$	674	W	29	29
Nimocinol	—	—	$C_{26}H_{32}O_2$	408	F	70	70
Nimolicinoic acid	52	92	$C_{26}H_{34}O_6$	442	F	36	36
Nimolicinol	45	270	$C_{28}H_{34}O_7$	482	S	33	33
Nimolinone	62	—	$C_{30}H_{44}O_3$	452	F	63	63
Salannin	4	167	$C_{34}H_{44}O_9$	596	S	28	28
3-Desacetyl-	29	214	$C_{32}H_{42}O_8$	554	S	16	16
Photooxidized-	46	244	$C_{34}H_{44}O_{10}$	612	S	13	13
Salannol	40	208	$C_{32}H_{44}O_8$	556	S	16	16
Scopoletin	—	204	$C_{10}H_8O_4$	192	T	65	65
Sugiol	2a	—	$C_{20}H_{28}O_2$	300	B	62	62
7-Acetylneotrichilenone	14	208	$C_{28}H_{36}O_5$	452	S	15	15

Table 1 (continued)
SUBSTANCES ISOLATED FROM VARIOUS PARTS OF THE NEEM TREE

Name	Structure	M.P.	Mol. Formula	Mol. Wt.	Source[a]	Isol.	Ref.
Vepinin	21	—	$C_{28}H_{36}O_5$	452	S	18	18
Vilasinin	18	255	$C_{26}H_{36}O_5$	428	L	16	16
3-Acetyl-7-tigloyl lactone-	50	242	$C_{33}H_{46}O_8$	570	S	35	35
1,3-Diacetyl-	20	157	$C_{30}H_{40}O_7$	512	S	16	60

[a] S = seeds, W = wood, L = leaves, B = bark, and F = fruit.

of the side chain are cyclized into a furan ring are called limonoids. Other changes may involve opening of one of the rings or rearrangement of the skeleton. Different plants tend to produce a range of products with one of these alterations. Those from neem with further skeletal changes tend to have the C-ring opened by oxidation, though examples of A-ring and D-ring opened products are known. Many of the oxygenation products can be seen to arise from formal epoxidations or Baeyer-Villiger reactions. The most important cleavage reaction for neem triterpenoids is the C-12-C-13 fission. In most compounds this means there is a C-12 carbomethoxy group, as in azadirachtin (Structure 1).

In the seeds, at least, it would appear that these successively more oxidized products are formed as ripening proceeds. The unripe fruit and the seeds inside them contain triterpenoids, but salannin (Structure 4) is the most highly oxidized product isolatable in quantity. As ripening proceeds, other products appear, and in fully ripened seeds, the most highly altered group of compounds, yet identified, represented by azadirachtin, rise to a maximum concentration (Morgan, unpublished results).

A. INTACT TRITERPENES

A number of oxidized intact triterpenes can be isolated from other species of Meliaceae. These so-called protolimonoids appear to be biosynthetic precursors of the limonoids (see next section) because of their biological occurrence and oxidation pattern. The only protolimonoids isolated from the neem tree are meliantriol (Structure 5), isolated by Lavie et al.[7] and nimbocinone (Structure 6), recently extracted by Siddiqui et al.[8] The preparatory steps towards the formation of the furan ring are evident in both compounds.

B. LIMONOIDS
1. With Ring System Intact

Limonoids are usually defined as triterpene derivatives from which four side-chain carbon atoms have been lost and the remainder have been cyclized to a furan ring, hence the alternative name tetranortriterpenoids. As well as the changes in the side chain, they also show a rearranged carbon skeleton, with migration of the double bond to ring D and the C-30 methyl group from C-14 to C-8. This rearrangement has been demonstrated *in vitro* by the conversion of turraeanthin (Figure 1a) from *Turraeanthus africanus* into a simple limonoid via a $7\alpha,8\alpha$-epoxide (Figure 1b) followed by rearrangement to a 7α-hydroxy-Δ^{14}-apo-derivative (Figure 1c) (compare Structures 7 and 8).

Lavie et al.[11a,11b] have isolated a number of simple limonoids from neem which are still closely related to the tetracyclic triterpenoids, such as azadirone (Structure 7), azadiradione (Structure 8), and epoxyazadiradione (Structure 10) also known as nimbinin. Kraus and Cramer[12] have isolated azadiradione 7-benzoate (Structure 9) and epoxyazadiradione benzoate (Structure 11).

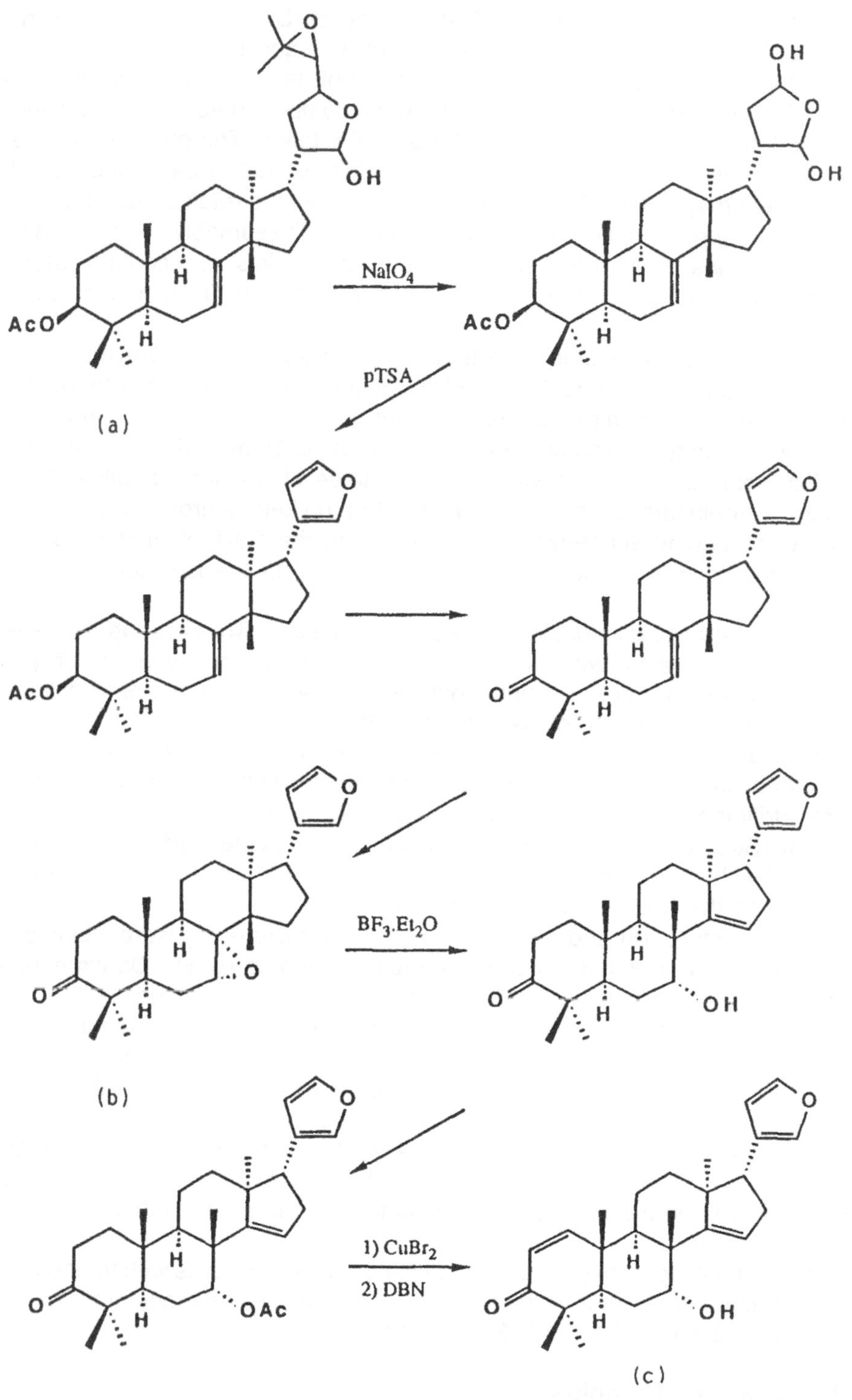

FIGURE 1.

The compound (Structure 10) was first isolated by Siddiqui,[4] and named nimbinin and later isolated independently by Connolly et al.[13] and by Narayanan et al.[14] with physical and spectroscopic data in good agreement and was assigned this structure. Lavie and Jain assigned the same structure to the compound epoxyazadiradione which they isolated, the only difference being the optical rotation. The material of Narayanan et al. and Connolly et al. give very similar rotations but that of Lavie and Jain differed both in sign and magnitude. However in a later full paper the rotation reported by Lavie's group had changed dramatically with no explanation. Presumably the explanation was that all four groups isolated the same material but Lavie's was contaminated. Lavie converted azadirone (Structure 8) to azadiradione (Structure 9) by a selenium dioxide oxidation.

Kraus and Cramer were able to determine the position of the β-hydroxyl, in 17-β-hydroxyazadiradione (Structure 12) by ^{13}C NMR spectral comparison with azadiradione (Structure 7), the C-17 peak becomes a singlet and shifts downfield from δ = 60.7(d) to 80.7(s) ppm. The furan was proved to be *cis* to the C-18 methyl by nOe experiments. Kraus also isolated one of the very rare examples of a limonoid with a 17-β furan substituent, 17-epiazadiradione (Structure 13), but the ketone group at C-16 makes the C-17 position open to epimerization. The isolation therefore of epimers about C-17 should be treated with caution, as they may arise from the isolation and extraction procedure.

Three novel tetracyclic limonoids were characterized by Kraus' group:[15] 7-acetylneotrichilenone (Structure 14) which contains an unusual 15-carbonyl group, diepoxyazadiradione (Structure 15) in which both conjugated double bonds are epoxidized, and 1-α-methoxy-1,2-dihydroazadiradione (Structure 16).

All these natural products were characterized using ^{13}C, ^{1}H, and ^{1}H nOe NMR techniques. 7-Acetylneotrichilenone (Structure 14) was crystallized from methanol and the structure additionally determined by X-ray crystallography.

Meldenin (Structure 17) isolated from neem and characterized by Connolly et al.[13] contains a 6-α-hydroxy and is presumably an intermediate compound between the azadirones and the C-6,C-28 ether linked limonoids.

Vilasinin (Structure 18) and vilasinin triacetate (Structure 19) were isolated by Narayanan's group[16] and the stereochemistry fully confirmed using nOe experiments by Kraus and Cramer[17] who also isolated vilasinin-1,3-diacetate (Structure 20). Both are related to the more complex salannin (Structure 4) (in which the C ring has opened) by the presence of a 28,6-α-ether linkage as well as the α,α-1,3-diol system.

Vepinin (Structure 21) isolated and characterized by Narayanan[18] also appears to provide a link between the tetracyclic azadirone type structures and the C-7,C-15 ether bridged compounds like salannin (Structure 4) and nimbin (Structure 22). The absence of a hydroxyl group in the NMR together with the chemical shifts of the H-7 and H-15 protons provided some evidence for the structure, but it has not otherwise been completely characterized.

Finally, the most highly oxygenated example of these limonoids with the nuclear rings still intact is azadirachtanin A (Structure 23), which as well as many other oxygenated functional groups has a C-19 to C-29 ether bridge.[19]

2. C-Ring Opened Limonoids

The most characteristic group of triterpene-derived compounds so far identified in neem are those limonoids in which ring C has been opened by oxidation, and it is in this group of compounds that the interesting biological properties begin to appear.

The first limonoid to be purified from *A. indica*, nimbin (Structure 22) — isolated by Siddiqui[4] in 1942 — belongs to this group. Nimbin is present in many parts of the tree;

Siddiqui isolated it as a crystalline solid from seed oil, blossoms, root, and trunk bark, but it took a further 20 years for its structure to be correctly determined. There was much debate on the molecular formula,[20] Narasimhan[21] correctly suggesting $C_{30}H_{36}O_9$, but although eight of the nine oxygens could be identified as belonging to two methyl esters, an acetate and α,β-unsaturated ketone and a β-substituted furan, the nature of the remaining oxygen was open to interpretation. Chemical studies initiated by Mitra[22] involved the hydrolysis of all three ester moieties to give deacetylnimbin (Structure 24), nimbinic acid (Structure 25), and nimbic acid (Structure 26). Narasimhan observed that nimbin could be selectively hydrogenated at the C-2,C-3 double bond and also that nimbic acid, upon sublimation, lost carbon dioxide and water to give an enol-lactone pyronimbic acid (Structure 27).[21]

Sengupta et al.[20] established a part structure (Structure 28) by analysis of the products resulting from loss of water and CO_2 from nimbic acid (Structure 26) in basic conditions. The acetoxy group was shown to be derived from a secondary alcohol. By the use of NMR and assuming that nimbin was a triterpene like other bitter principles, Narayanan's group solved the greater part of the structure.[23] The "final" oxygen was identified as being an ether bridge between C-7 and C-15. The ring D structure was determined by observation of the distinctive C-13 methyl to C-15 long range coupling and the C-15, C-16, and C-17 proton couplings in the NMR spectrum. Narayanan[24] further explained the origin of their stereochemical assignments in a second paper and noted the distinctive axial-axial-equatorial couplings of the C-5, C-6, and C-7 protons. The absolute stereochemistry was assigned on the basis of the similarity of the ORD curve of hexahydronimbin to that of 1-ketocholestane. Unfortunately, the configurations at C-4 and C-15 were incorrect. These were corrected with contributions from Overton's group[25] who assigned the methyl group at C-4 to the β-configuration because of the ease of lactone formation from the C-28 carboxylate to the C-6 hydroxyl group in deacetylnimbin (Structure 24). Also the proton at C-15 was assigned the α-configuration by analogy with decoupling experiments on salannin (Structure 4). These corrections were later confirmed by Narayanan and Pachapurkar.[26,27]

Overton's group also solved the structure of salannin (Structure 4).[28] With the help of NMR, the similarities to nimbin (Structure 22) were obvious, notably the β-substituted furan, the ring D pattern, and the characteristic C-5, C-6, and C-7 proton patterns. However only a single methyl ester was present, and in the NMR spectrum there were two characteristic doublets at approximately $\delta = 5$ ppm as well as tiglate ester signals in place of the ring A enone resonances. This data led to the Structure 4; the placing of the tiglate ester at the C-1 position was determined by observing shift differences at the C-12 methyl in salannin, desacetylsalannin (Structure 29), and desacetyldetigloyl-salannin (Structure 30).

Ekong et al.[29] isolated two new limonoids from the trunk wood of *A. indica*. Nimbolin A (Structure 31) was characterized by ^{1}H NMR spectroscopic comparisons to azadiradione (Structure 8) and salannin (Structure 4). Nimbolin B (Structure 32) was also characterized and its structure, in which the C ring has opened, determined.

A proposed mechanism for the formation of nimbolin B from a hypothetical intermediate (Structure 33) is shown in Figure 2. This could prove to be a plausible mechanism for the biosynthetic transformations of tetracyclic limonoids to C-ring opened compounds of the salannin type, but no biosynthetic studies in this area have been carried out.

Mitra[22] isolated the natural product nimbidinin (Structure 34) confirming its structure using ^{1}H NMR, infrared, and extensive mass spectrometry. There is a similarity to salannin and a proposed route to the C-ring opened limonoids is shown in Figure 3.

Many of the compounds described in this section have closely related structures, so

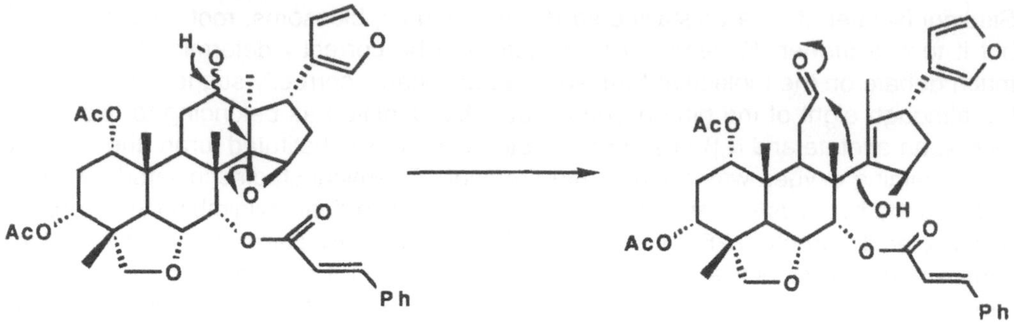

(33)

FIGURE 2.

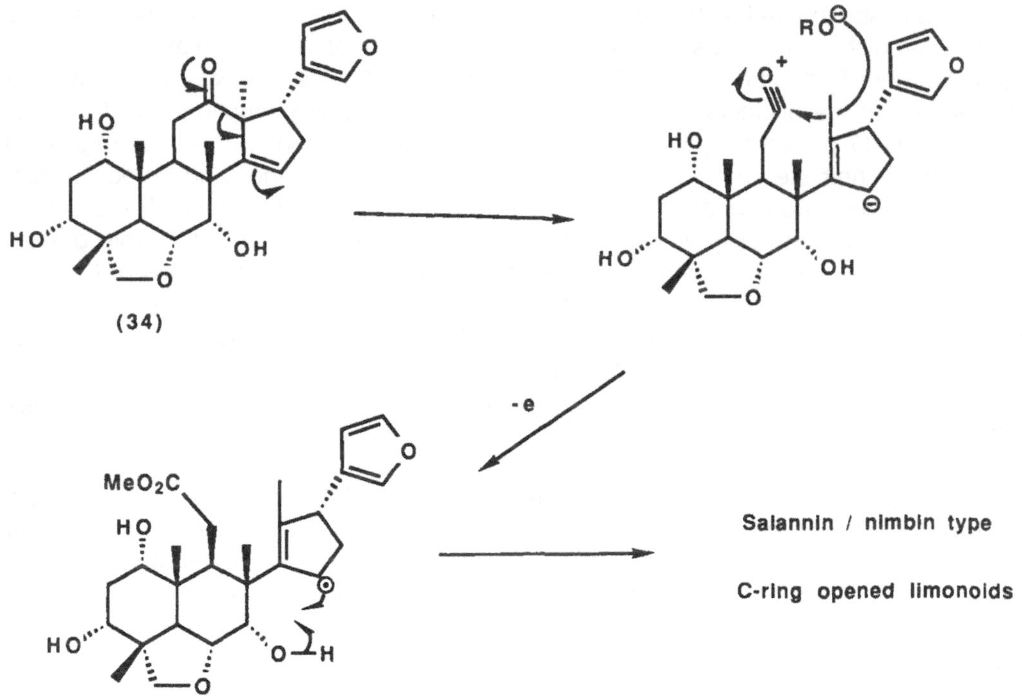

(34)

-e

Salannin / nimbin type

C-ring opened limonoids

FIGURE 3.

later identifications were based on earlier structure solutions. Inevitably this has caused problems and some structures are doubtless incorrect. An example is nimbolide (Structure 35) isolated in 1967.[30] It was shown, by NMR, to be very closely related to nimbin (Structure 22); also the dihydro-derivative was converted to 2,3-dihydronimbic acid which had been prepared by Sengupta earlier.[20] A problem arises with the configuration at C-15. The β-H configuration is claimed, presumably on the basis of the configuration originally proposed by Narayanan et al.[24] There seems little doubt that the α-H configuration (as drawn in Structure 35) is correct.

Several more limonoids isolated by Kraus and Cramer[31] contain the nimbin type structure. Nimbinene (Structure 36) and deacetylnimbinene (Structure 37) are both pentanorterpenoids in which the β-carboxylate has been lost leaving a 3,4-double bond. Desacetylsalannin (Structure 29) was found to be a natural product. Also isolated were nimbandiol (Structure 38) and 6-acetylnimbandiol (Structure 39) which have lost a carboxylate and been reoxidized at C-4. Kraus and Cramer[17] also report the isolation of salannol (Structure 40) a close derivative of salannin (Structure 4).

3. A-Ring Opened Limonoids

The only example of an A-ring opened limonoid from the neem tree is $4\alpha,6\alpha$-dihydroxy-A-homoazadirone (Structure 41) isolated by Kraus' group.[32] The A-ring has undergone a ring expansion, between C-3 and C-4 to a seven-membered ring.

4. D-Ring Opened Limonoids

Gedunin (Structure 42) is a D-ring opened limonoid first isolated from *Entandrophragma angolense* which has also been found in *A. indica*.[11,15] Kraus et al.[15] have also isolated the 7-benzoyl derivative (Structure 43) of gedunin. Lavie et al.[11a,11b] also report that 7-desacetylgedunin (Structure 44) is present in *A. indica* and that it is possible to convert epoxyazadiradione (Structure 10) to gedunin (Structure 42) by Baeyer-Villiger oxidation, possibly mimicking a biosynthetic conversion.

Nimolicinol (Structure 45) is only the second example (see Structure 13) of a limonoid containing a β-linked furan. Nimolicinol is probably derived from an azadiradione type compound. The workers in this case were particular in stating that the isolation conditions were mild, so as to avoid epimerizations.[33] However, the lactone ring hides a ketone at C-17 which makes epimerization easy.

C. PRODUCTS WITHOUT A FURAN RING

There are a number of examples now known which would otherwise be classed as limonoids, but which do not contain the furan ring. The first such compounds contained C-20,C-23 lactones, regarded as photooxidation products, possibly formed during isolations, e.g., the nimbin and salannin derivatives (Structures 46 and 47), respectively.

Most recently two examples have been discovered which contain the C-17 salannin furan-substituent in the form of an α,β-unsaturated lactam carbonyl.[34] Salannolactam I (Structure 48) and II (Structure 49) differ only in the position of the lactam carbonyl group. A vilasinnin derivative, 3-acetyl-7-tigloylvilasinnin lactone (Structure 50) has recently been isolated by Ley's group and the structure confirmed by X-ray crystallography.[35] Two further nonfuranoid compounds, isonimolicinolide (Structure 51) and the hexanortriterpenoid, nimolicinoic acid (Structure 52) have recently been isolated from neutral fruit extracts,[36] and yet another isoazadirolide (Structure 53) from leaves.[37]

1. Azadirachtin and Related Compounds

Azadirachtin (Structure 1), from a commercial and biological point of view is the most interesting of the tetranortriterpenoids from neem. It was discovered in 1967 by Morgan by following antifeedant activity for the desert locust *Schistocerca gregaria* in whole seed extracts. By column and thin layer chromatography a colorless, microcrystalline solid was isolated with remarkable antifeedant activity of 70 µg/l for the desert locust.[1] The molecular formula $C_{35}H_{44}O_{16}$ was determined by accurate mass measurement on a trimethylsilyl derivative.[38] Morgan recognized the importance of the discovery and extended the work to identify the functional groups, to classical degradation methods, and attempts to find a truly crystalline derivative. Two methyl esters, two acetates, a tiglate ester, a secondary and two tertiary hydroxyls, and a dihydrofuran ring linked to another ether ring were identified chiefly from NMR and mass spectral information.[39] In fact, one of the tertiary alcohols and one ether linkage are a hemiacetal, and one of the acetates was incorrectly identified, but is in fact a quarternary methyl with its NMR absorption shifted to unusually low field ($\delta 2$). Attempts to demonstrate the presence of an epoxide were all negative. The remaining two oxygens are present in ether and epoxide linkages, but it took another 15 years to elucidate the correct structure. A monoacetate (of one of the tertiary hydroxyls) and various trimethylsilyl ethers, up to a tris(trimethylsilyl)ether, were made but the resistance of the secondary hydroxyl to acetylation or oxidation suggested it was very hindered. Low field NMR doublets at δ6.4 and δ5.05 were correctly assigned to a dihydrofuran ring and a singlet at δ5.64 to the other dihydrofuran proton. The lack of further couplings gave the part Structure 54. The part Structure 55 was derived from NMR studies, selenium dehydrogenation, and removal of an acetate and tiglate group. Other assignments (including that this compound belonged to the C-seco limonoid series) were unacceptable at the time to referees on the grounds of insufficient evidence. A major advance had, however, been made in identifying the structural units in this important compound.

Further work with 100 MHz NMR and ^{1}H-^{1}H decoupling experiments did not advance the work much because the many oxygen atoms and the quaternary carbon atoms prevent extensive proton coupling. Nakanishi (see Zanno et al.[40]) in 1975 made a major advance by the application of newer NMR methods, notably ^{13}C studies. By comparison with the ^{13}C spectra of salannin and nimbin, all 35 carbon atoms were assigned as in Structure 56. The very low chemical shift of the 18-methyl was attributed to its *cis* relationship to the C-20 hydroxyl group. A long range coupling between the protons of the C-10 methyl and H-5 could be observed due to their *trans* peri-planar arrangement.

However, it was evident there were errors in the Nakanishi structure. Renewed assaults were made on the problem by groups led by Ley,[41,42] Kraus,[43,44] and Nakanishi,[45] through NMR methods and revived efforts to find a crystalline derivative. Further NMR work showed that the ether bridge from C-11 was connected to C-19 as opposed to C-30. Kraus showed by the use of deuterium exchange experiments that the second tertiary hydroxyl group was attached to C-11, making a hemiacetal moiety.[41] The chemical shifts of C-13 and C-14 in the carbon NMR spectra could now be explained by an epoxide link. Rotation around the C-8, C-14 could now be expected, but strong hydrogen bonds were shown to exist between the "two halves" in the X-ray structure of the detigloyldihydro-derivative published by Ley (see Broughton et al.[42]). The restricted rotation was clearly observed in a low temperature NMR experiment which showed doubling of virtually all the proton signals.[45]

X-ray crystallography was an obvious method for the unambiguous solution of the structure of azadirachtin, but the preparation of crystals had eluded most workers. It was only during the preparation of simple derivatives as part of a structure-activity study that a derivative, first made by Morgan (see Butterworth et al.[39]) was re-prepared and obtained crystalline.[42]

Since the correct structure of azadirachtin was published, several more azadirachtinoid structures have been isolated and are listed in Table 1. Of special interest is 1-tigloyl-3-acetyl-11-methoxyazadirachtinin (Structure 57) isolated by Kraus et al.[44] This compound formally results from the nucleophilic opening of the epoxide by the C-7 hydroxyl group, with Walden inversion. Some more azadirachtin-like limonoids are 3-deacetyl-3-cinnamoylazadirachtin[44] (Structure 58), 22,23-dihydro-23-β-methoxyazadirachtin[43] (Structure 59), and 3-tigloylazadirachtol[35,44] (Structure 60), and some structures still remain doubtful.[46]

As a final note on structure determination, several of the compounds in Table 1 were "solved" before the advent of high field NMR, and while their gross structures are in little doubt, those with epimerizable centers, notably the azadiradiones, may contain errors. ^{1}H NMR data for a number of representative neem compounds are collected in Table 2, and ^{13}C data are in Table 3.

Table 2
¹H NMR DATA ON SOME SELECTED NEEM COMPOUNDS

H	Azadirachtin (1)	1-Tigloyl-3-acetyl-11-methoxy azadirachtinin (57)	Azadiradione (8)	Dihydroxy-homoazadiradione (41)	Meldenin (17)	Nimbin (22)	Salannin (4)	Vepinin (21)	Vilasinin lactone (50)
1	4.75 dd[a]	4.81dd	7.17d	6.66d	nq[b]	—	4.79t	7.05d	3.57t
2	2.34ddd	2.13ddd				5.89d			2.01m
2	2.13ddd	2.28ddd	5.83d	5.93dd	nq		2.38m	5.73d	2.33dt
3	3.50dd	5.48dd				6.35d	4.96t		5.10t
4	—	—	—	—	—	—	—	—	—
5	5.35d	3.16d	nq	2.33d		2.99m	2.80d	nq	2.41d
6	4.60d	4.39ddd	nq	4.50ddd	2.14q	5.21dd	3.99dd	5.28dd	4.18dd
7	4.75d	4.53d	5.32t	5.31d	4.09d	4.05d	4.18d	4.18d	5.63d
9	3.34s	3.56s	nq	2.21m	nq	2.99m	2.75dd	nq	2.79dd
11	—	—	nq	1.7-2m	nq	2.30	2.38m	nq	1.80m
12	—	—	nq	1.7-2m	nq	—	—	nq	1.80m
14	—	—	—	—	—	—	—	—	—
15	4.67d	4.13m	5.83s	5.46d	5.58	5.57t	5.43t	4.05m	5.43dd
16	1.73	1.85		2.41ddd	nq	2.30m	2.38m	nq	
16	1.31d	2.15m		2.36ddd	nq	3.71m	3.72m	nq	2.13m
17	2.38d	2.12m	3.43s	2.85dd		1.69d	1.68s	1.22*	nq
18	2.01s	1.50s	1.03s*[c]	0.79s	1.28*	1.29s*	1.21s	1.19*	1.13s
19	3.63d	3.73d	1.08*	1.27s	1.22*				0.96s
19	4.15d	4.21d							2.75m
20	—	—	—	—	—	—	—	—	3.91t
21	5.65s	5.64s	7.45	7.39	7.38*	7.22s*	7.27s*	7.30*	4.45t
22	—	4.88d	6.28m	6.28m	6.30*	6.35s	6.29s	6.21*	2.20dd
23	5.05d	6.39d	nq	7.25m	7.26*	7.32t*	7.33t*	7.20*	2.50dd
28	6.46d	3.66d	1.08s*	2.65dd	1.10*			1.04*	3.50d
28	4.08d	4.03d	1.19s*	3.16dd			3.72m	1.04*	3.27d
29	3.76d	—	1.35s*	1.56s	0.88*	1.36s*	0.98s*		1.19s
30	1.74s	1.57s		1.31s	0.83*	1.37s*	1.29s*	0.8*	1.02s

[a] d = doublet, dd = doublet of doublets, ddd = doublet of doublets of doublets, t = triplet, m = multiplet, s = singlet, q = quartet.

[b] nq = Data not quoted.

[c] * = Assignment to a type of proton, e.g., to quaternary methyl or furan, not necessarily precise.

Table 3
^{13}C NMR DATA FOR SELECTED COMPOUNDS FROM NEEM

C	Azadirachtin (1)	1-Tigloyl-3-acetyl-11-methoxy azadirachtinin (57)	Dihydroxy-homoazadiradione (41)	Deacetylsalannin (29)	Acetylneotrichilenone (14)	Diacetylvilasinin (20)
1	70.51d	70.26d	154.35d	72.74d	157.83d	72.28d
2	29.37t	30.15t	128.86d	30.42t	125.98d	27.66t
3	66.99d	66.97d	199.86s	70.66d	204.41s	71.76d
4	52.52s	52.87s	74.99s	44.11s	44.11s	42.32s
5	37.06d	36.34d	53.71d	38.71d	45.15d	39.62d
6	73.79d	71.93d	68.52d	72.35d	22.26t	72.90d
7	74.37d	82.70d	78.58d	85.77d	73.36d	74.01d
8	45.41s	51.46s	43.86s	48.88s	41.08s	45.80s
9	44.69d	47.92d	39.67d	39.39d	47.10d	33.62d
10	50.19s	49.78s	42.25s	40.86s	39.59s	39.20s
11	104.10s	107.25s	18.48t	30.42t	17.68t	15.21t
12	171.70s	170.14s	34.10t	172.52s	34.45t	33.02t
13	68.53s	95.10s	46.68s	134.75s	42.16s	47.36s
14	69.69s	93.26s	157.96s	146.49s	61.33d	159.88s
15	76.43d	81.42d	120.47d	87.76d	218.61s	120.75d
16	25.06t	29.63t	34.48t	41.18t	43.23d	34.35t
17	48.67d	50.81d	51.77d	49.31d	38.00d	51.58d
18	18.49q	26.67q	22.18q	12.97q	27.89q	21.12q
19	69.07t	70.44t	14.98q	15.08q	19.31q	26.20q
20	83.55s	86.44s	124.23s	127.05s	122.89s	124.52s
21	108.70d	109.41d	139.67d	138.65d	140.28d	139.76d
22	107.30d	108.24d	110.85d	110.54d	110.83d	111.06d
23	147.00d	nq	142.67d	142.75d	143.01d	142.62d
28	nq	nq	57.98t	77.71t	27.37q	77.88t
29	173.20s	173.34s	27.28q	19.79q	21.13q	15.37q
30	21.33q	17.63q	26.40q	16.77q	18.20q	19.47q

D. ISOLATION

Methods of isolation of triterpenoids depend upon the part of the tree chosen, but as the most familiar hunting ground for chemists is in the seed oil, some discussion of this material is useful.

The kinds and amounts of the various substances isolatable varies enormously. No doubt the biotype, climate, ripeness, harvest, and storage conditions all play a part but which are most important is certainly unknown. Unfortunately, no quantitative studies have been carried out, but from our own work it seems that the major substance recoverable varies from nimbin to salannin to azadirachtin as one moves from fruit to seeds of increasing ripeness.

Extraction of a kilogram of seeds (of undefined moisture content) with ethanol, methanol, or acetone yields 100 to 200 g of dark brown oil. Solvent partition of this between light petroleum and aqueous methanol removes the triglycerides and leaves in the methanolic phase a semisolid mass of triterpenoids (50 to 150 g). The amount of water used with the methanol can vary from 5% up to 35%, when a third insoluble phase begins to form, in order to control the amount of less polar substances extracted from the oil. Repeated washing with light petroleum to remove more nonpolar material or other solvent partitions can be used. Ethyl acetate-water partitioning is particularly useful to remove high polarity impurities in the aqueous phase. The residue in the ethyl acetate phase can then be submitted to any of a number of chromatographic methods.[38,47-49] Unfortunately many of the highly polar substances, such as azadirachtin are unstable or irreversibly adsorbed to some extent on activated alumina and silica,[39] and it has been observed that many limonoids can isomerise on silica gel.[49] Hydrolysis of esters can also occur, adding to the problems of reliable isolation. Octadecyl-coated silica does not cause adsorption but is expensive on a large scale and has poorer resolving power. Florex LVM, a grade of atapulgite clay, has been used with success as a column chromatography material,[38] but preparative reversed-phase HPLC is being increasingly used, particularly for the isolation of new and minor products.

E. SYNTHESIS

Several simple manipulations of the limonoids have been achieved during structure determinations, but it is perhaps not surprising that little synthetic work has been attempted on the difficult C-seco ring systems.

The conversion of turraeanthin to azadirone has been described (Figure 1). A curious rearrangement of salannin on treatment with trimethylsilyl chloride and sodium iodide gives a structure confirmed by X-ray crystallography (Structure 61, Figure 4).[51] Several modifications of the azadirachtin structure have been achieved as part of structural studies[39,42] or for looking at structure-activity relationships.[52] The trimethyl ether has been made for NMR studies.[45] The monoacetate[39] is now identified as C-11 acetoxy from COSY and nOe difference NMR experiments.[53] Hydrogenation can selectively saturate the dihydrofuran ring permitting selective removal of either acetate or tiglate. A rearrangement, possibly mimicking the biosynthetic pathway to the azadirachtin group has also been achieved on dihydroazadirachtin using either acidic or basic conditions, or an attempted deoxygenation of the epoxide (Figure 5).

There have been no attempts at a total synthesis of any of these neem limonoids but Ley's group has synthesized the D-ring-dihydrofuran portion (Figure 6) of azadirachtin and demonstrated its biological activity.[54] A simpler version of this portion of the molecule has also been synthesized by Muckensturm's group.[55] An approach to the decalin portion of azadirachtin has also been developed by Ley's group, giving a product with five of the seven chiral centers already in place (Figure 7).[56]

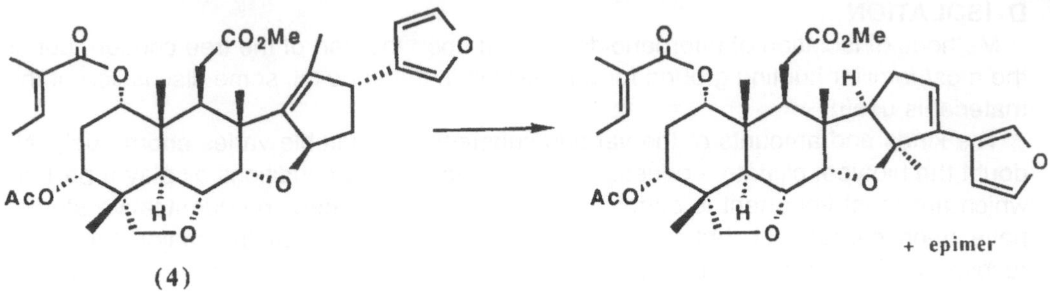

FIGURE 4.

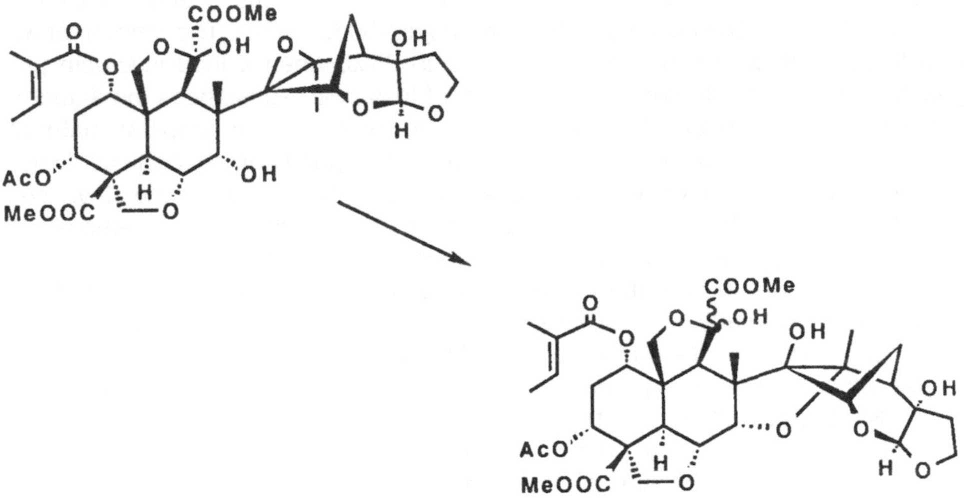

FIGURE 5.

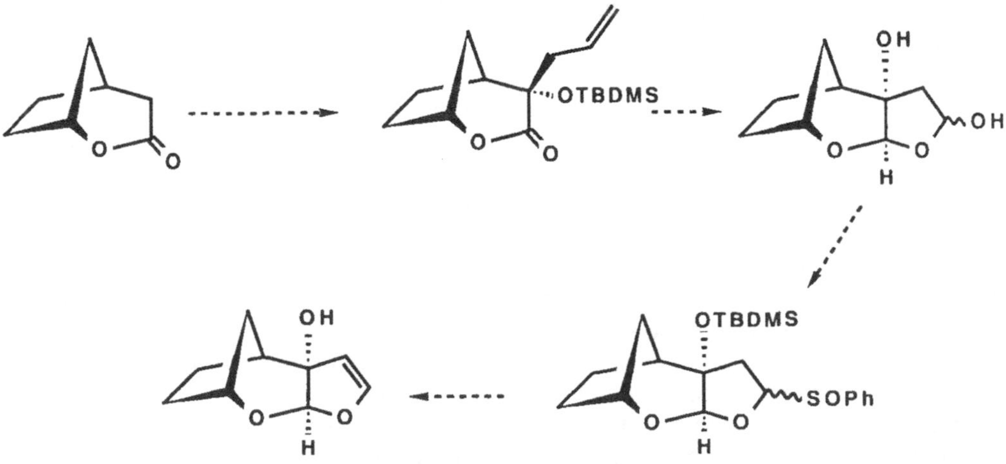

FIGURE 6.

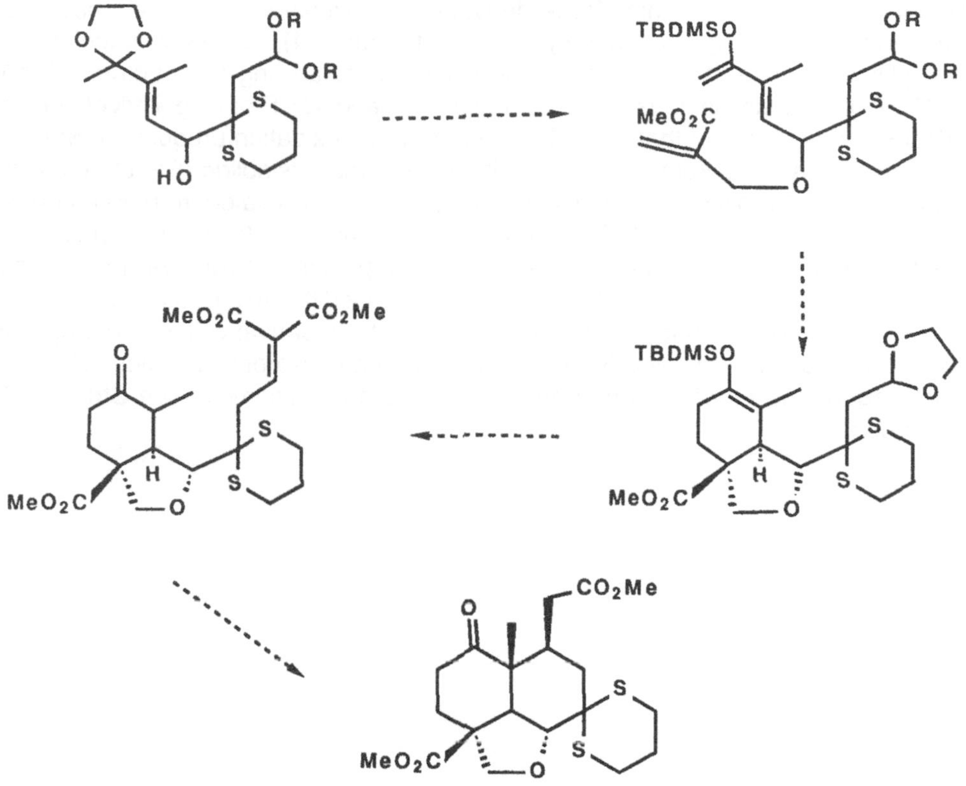

FIGURE 7.

III. CONCLUSION

There can be no other plant as the neem tree for which such a wide range of triterpenoid-derived substances is known, and certainly none that has been shown to produce such highly oxidized products. It is surprising therefore that no biosynthetic studies have been made, and the more-so in view of the importance of these compounds. The time seems opportune for an assault upon this subject, that may both enlighten us upon biochemical mechanisms in higher plants and perhaps enable us to understand and control the production of those useful substances in this plant.

Addendum

Since the preparation of this manuscript, further papers have continued to appear on the isolation of triterpenoids from neem, chiefly from Siddiqui's group. Authors have not displayed much restraint in their choice of new names for compounds and the whole subject has now become a minefield. A new triterpenoid with side chain intact (Structure 61) has been called azadirachtol,[62] a name which has otherwise been used for the 11-desoxy compound related to azadirachtin (Structure 60). We have amended the published 18-β-methyl structure to show an 18-α-methyl group in Structure 61 as more probable. Another triterpenoid with intact side chain, nimolinone (Structure 62) has been extracted from fruit.[63] Nimbocinol,[64] or rather 7-desacetyl-7α-hydroxyazadiradione (see Structure 8, with R = H) and isonimbocinolide[65] (Structure 63) are two further limonoids with ring structures intact. Note that nimbocinol and nimbocinone (Structure 6) have no particular structural relationship. Three derivatives of nimbin (Structure 22) have been isolated from green twigs: 6-desacetylnimbin (Structure 24), desacetylnimbolide, and desacetylisonimbolide, (two compounds in which the furan ring of Structure 24 is replaced by isomeric lactide rings,[66] as in Structures 51 and 53). From the work of Grimmer and Kraus,[67] it is probable that these lactides are photooxidation products of Structure 24. Two further lactides from the Seco-C-ring series, margosinolide (Structure 64) and isomargosinolide (with lactide carbonyl at C-23), have been isolated from green twigs.[68] Siddiqui's group has also reported, incidentally, the isolation of cholesterol from the fruit, of sitosterol and stigmasterol from leaves, and of scopoletin (7-hydroxy-6-methoxycoumarin) from the twigs. Klocke (see Lee et al.[69]) has trapped the volatile sulfur constituents from freshly ground seed and found they consist chiefly of derivatives of di-*n*-propyl and *n*-propyl-1-propenyldi-, tri-, and tetrasulfides, of which di-*n*-propyl disulfide (75%) was the most abundant. Finally, another group has isolated 4-epinimbin (see Structure 22) from seeds.[70]

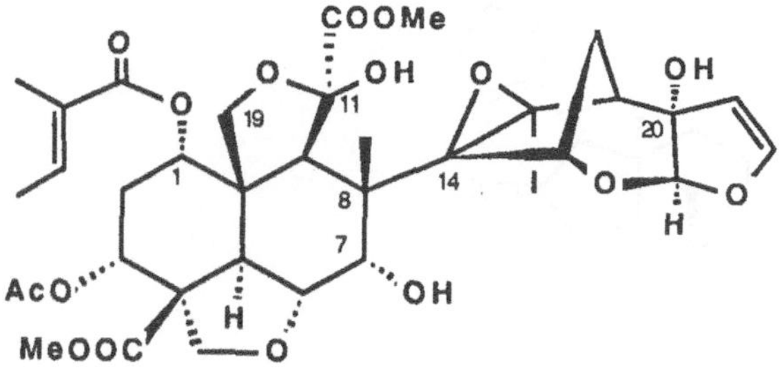

(1)   azadirachtin

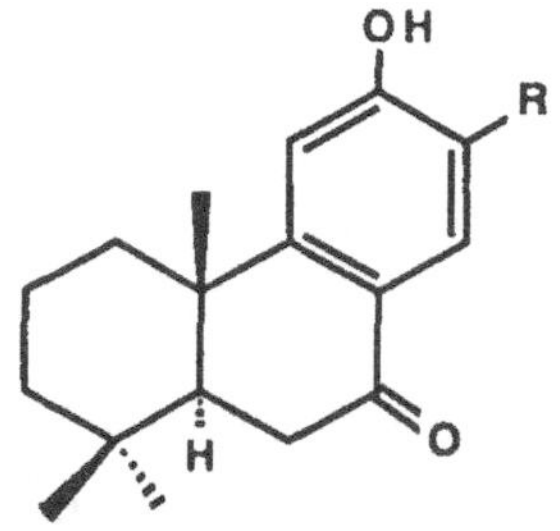

(2)   (a) sugiol   R= Me
      (b) nimbiol R= iPr

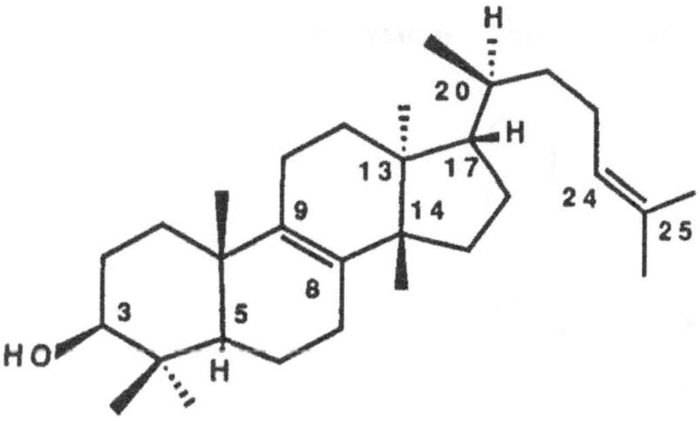

(3)   tirucallol

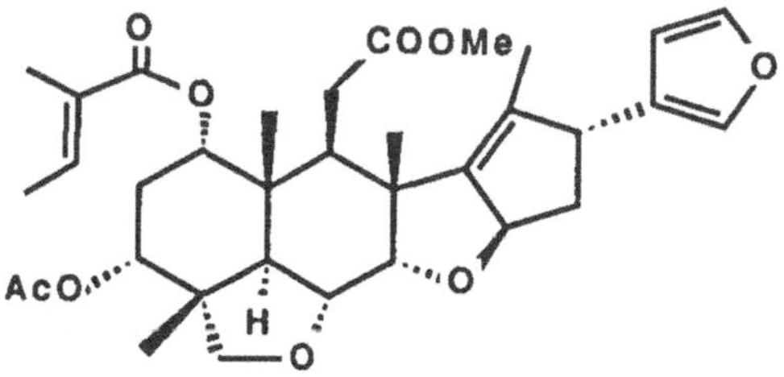

(4)   salannin

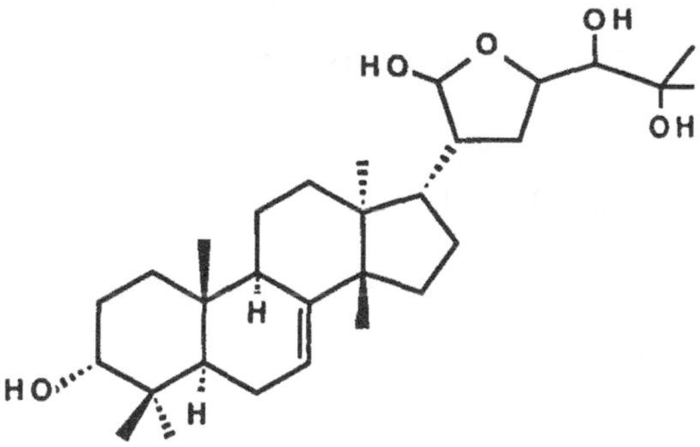

(5)   meliatriol

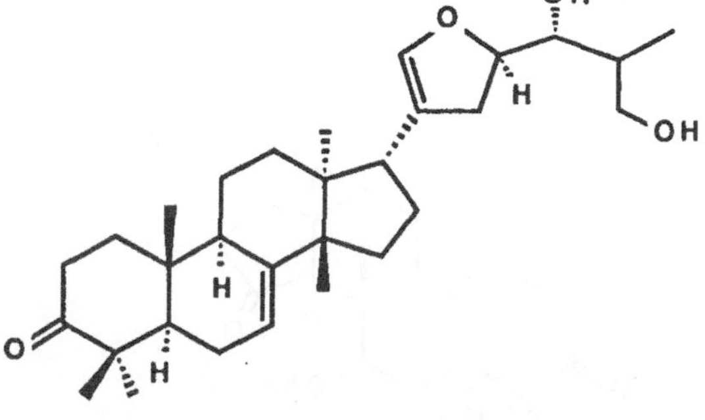

(6)   nimbocenone

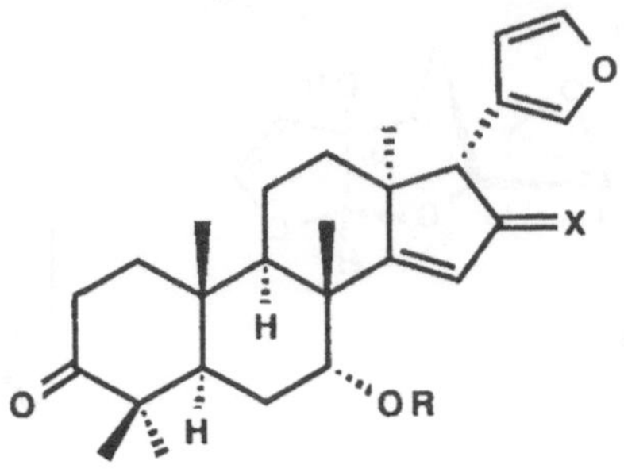

(7)   X=H,H   R=OAc   azadirone

(8)   X=O   R=Ac   azadiradione

(9)   X=O   R=COPh   azadiradione   benzoate

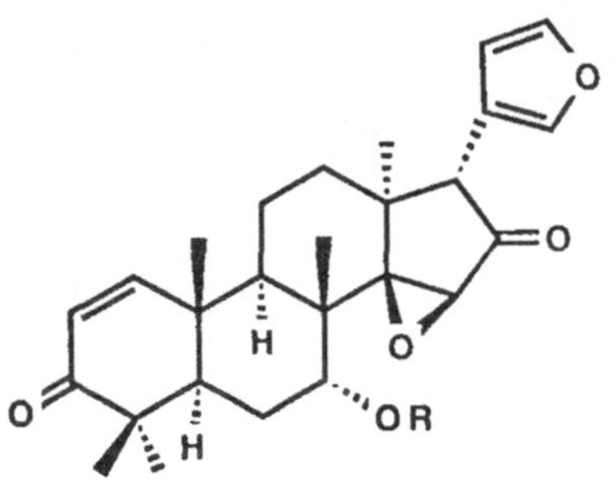

(10)   R=Ac   epoxyazadiradione

(11)   R=COPh   epoxyazadiradione   benzoate

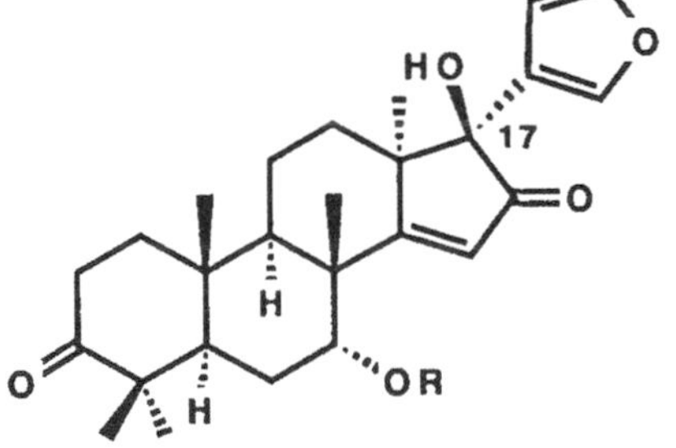

(12)   17-β - hydroxyazadiradione

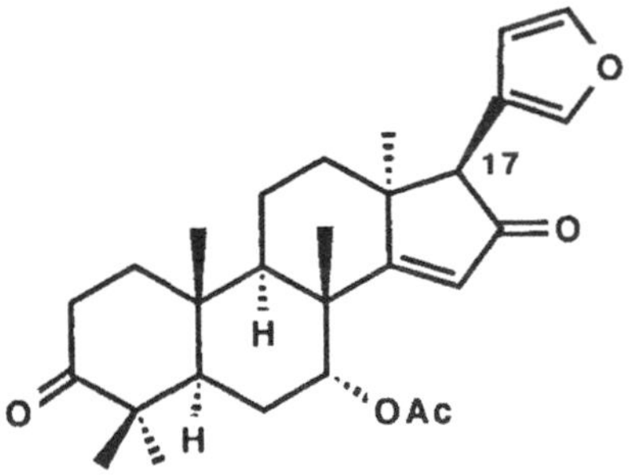

(13)   17-epiazadiradione

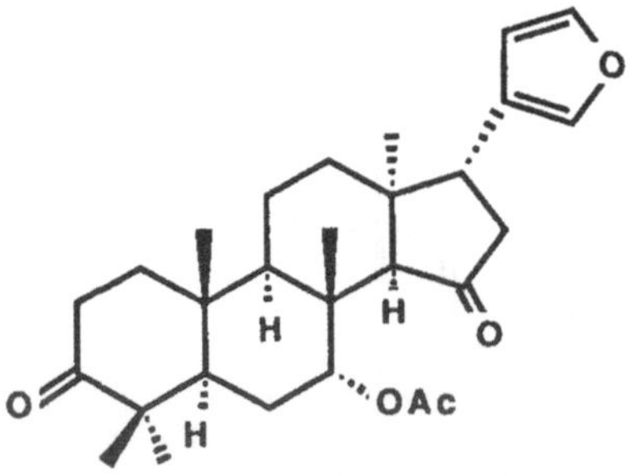

(14)   7-acetylneotrichilenone

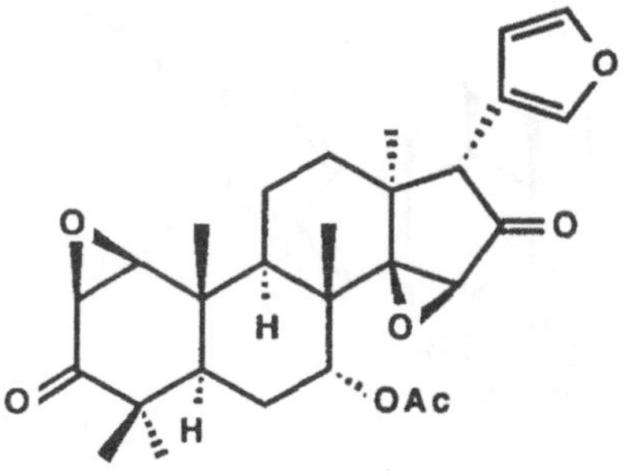

(15) diepoxyazadiradione

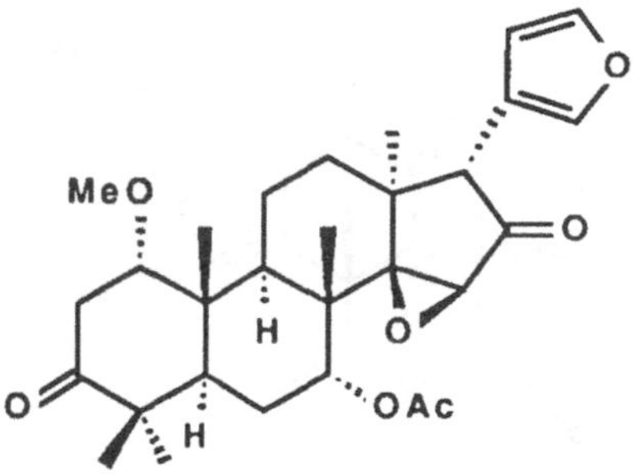

(16) 1-α-methoxy-1,2-dihydroazadiradione

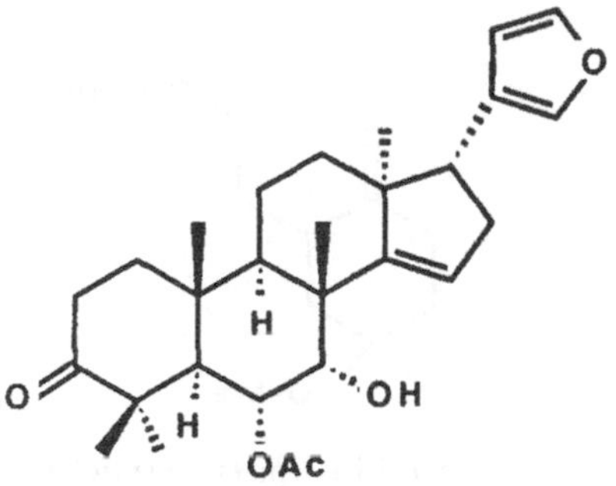

(17) meldenin

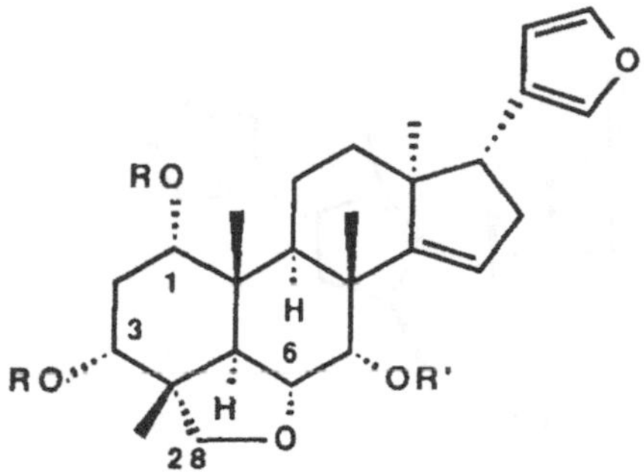

(18)   R=R'=H   vilasinin

(19)   R=R'=Ac   vilasinin triacetate

(20)   R=Ac,   R'=H   vilasinin-1,3-diacetate

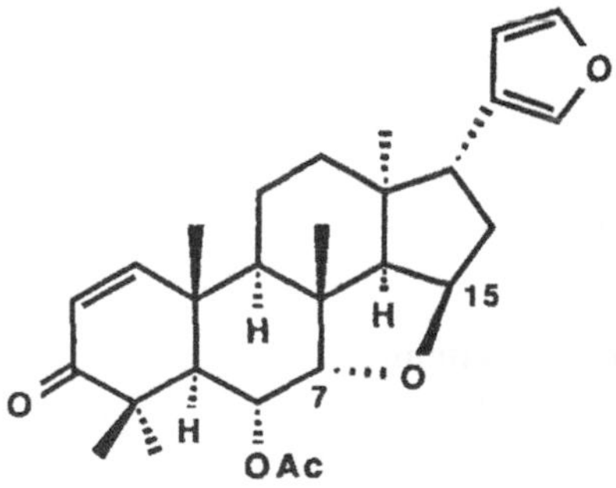

(21)   vepinin

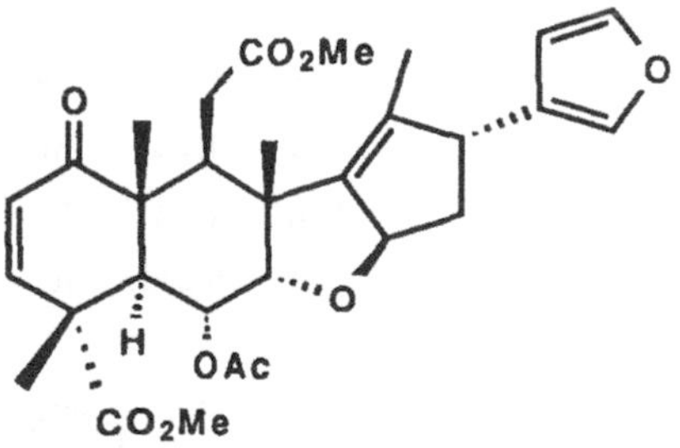

(22)   nimbin

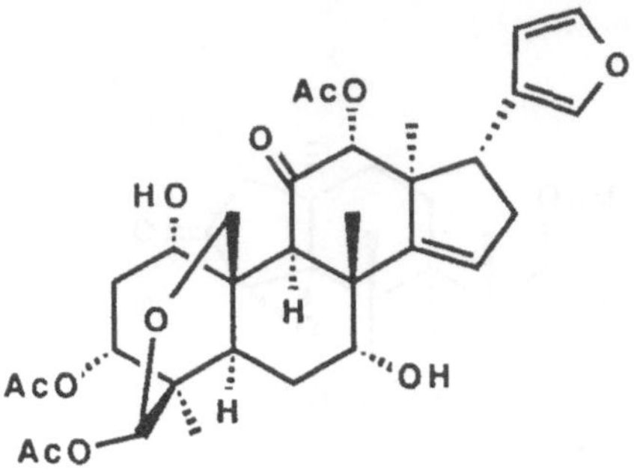

(23)   azadirachtanin A

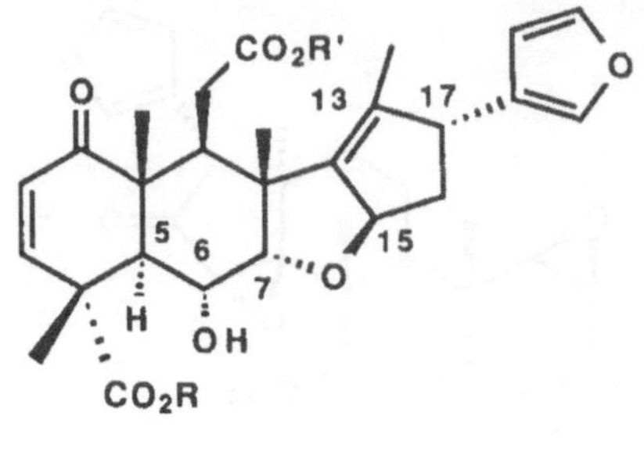

(24)   R=R'=Me   deacetylnimbin
(25)   R=Me, R'=H   nimbinic acid
(26)   R=R'=H   nimbic acid

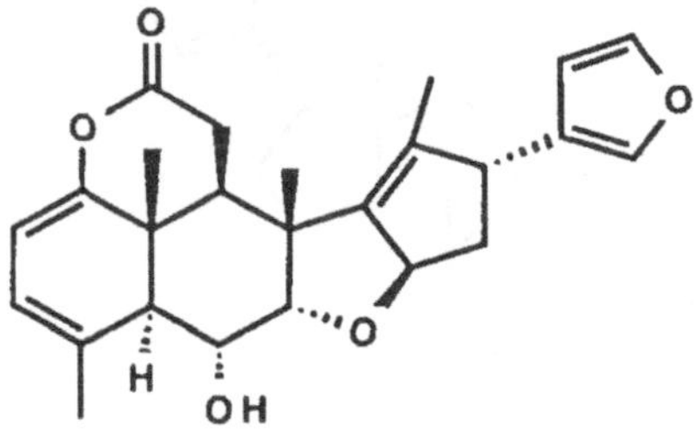

(27)   pyronimbic acid

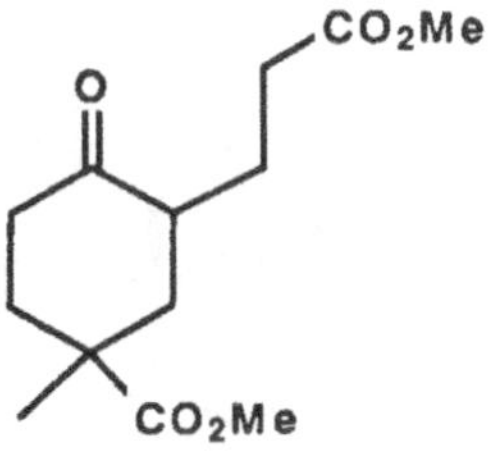

(28)   nimbin part structure

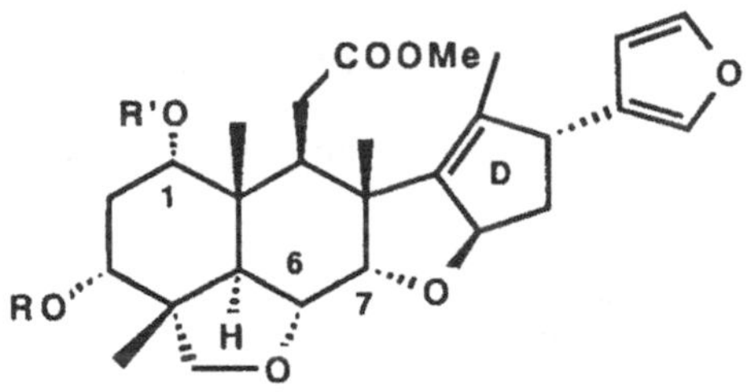

(4)   R=OAc, R'=tigloyl   salannin
(29)   R=H, R'=tigloyl   deacetylsalannin
(30)   R=R'=H   deacetylsalannin
(40)   R=H, R'=3-methylbutanoate   salannol

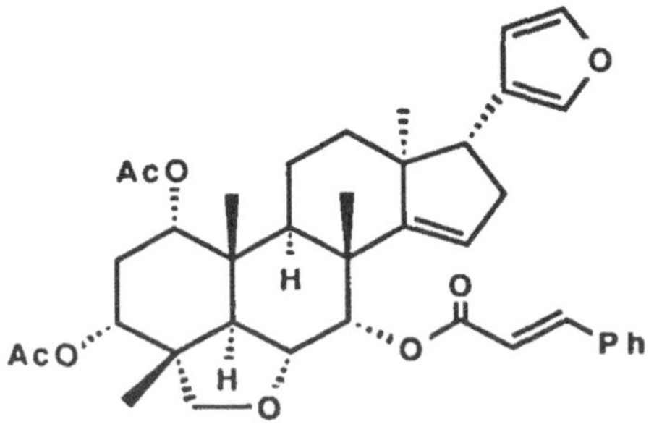

(31)   nimbolin A

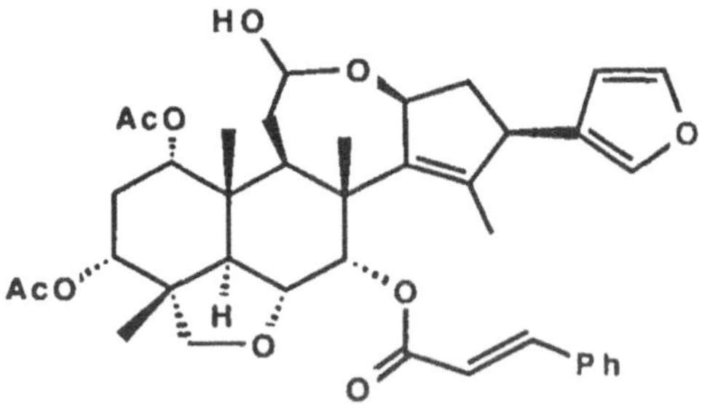

(32)   nimbolin B

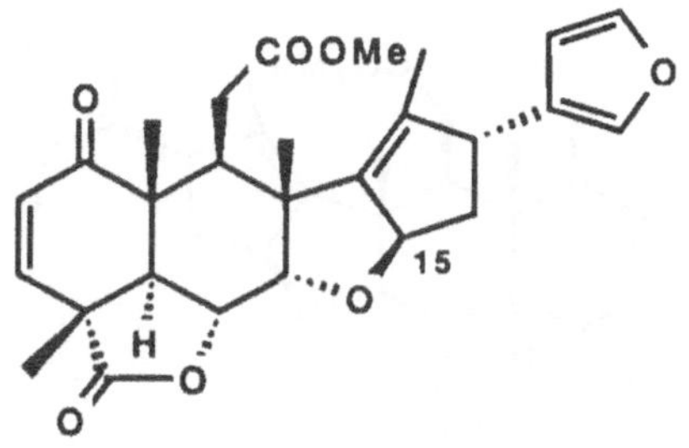

(35)   nimbolide

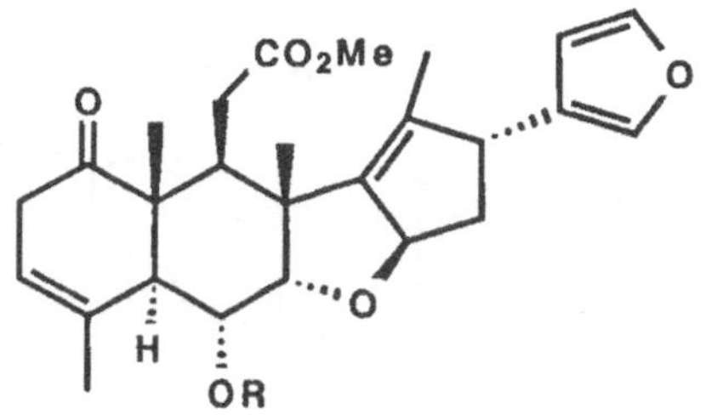

(36)   R=Ac   nimbinene
(37)   R=H   deacetylnimbinene

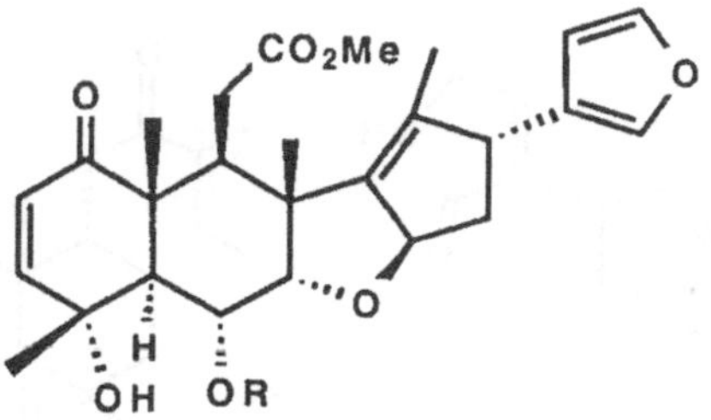

(38)   R=H   nimbandiol
(39)   R=Ac   6-acetylnimbandiol

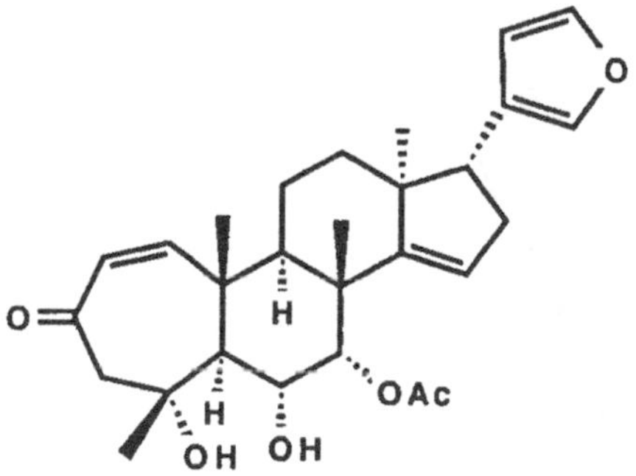

(41)   4α,6α   -dihydroxy-A-homoazadirone

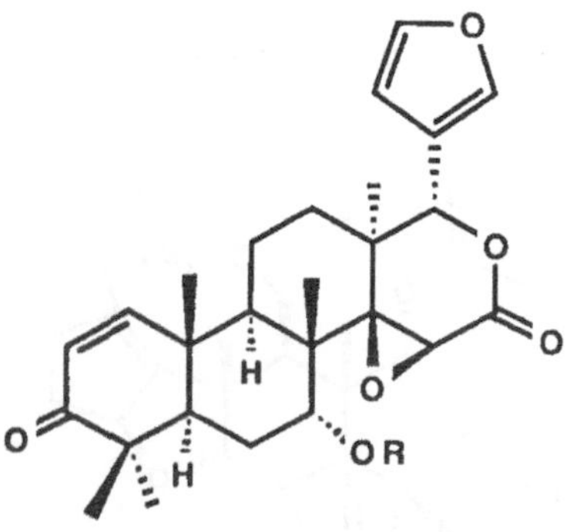

(42)   R=Ac   gedunin

(43)   R=COPh   7-deacetyl-7-benzoylgedunin

(44)   R=H   7-deacetylgedunin

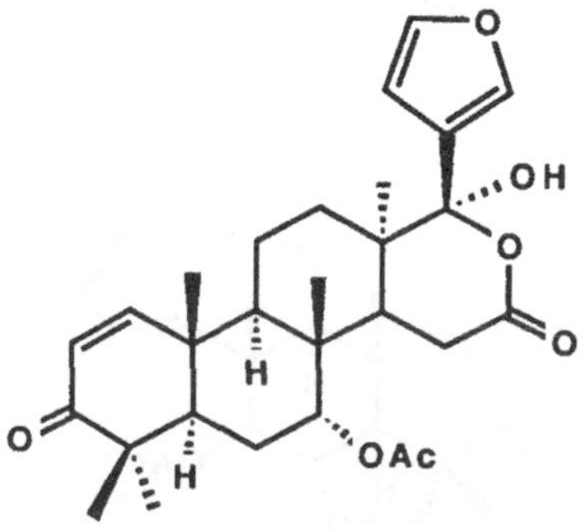

(45)   nimolicinol

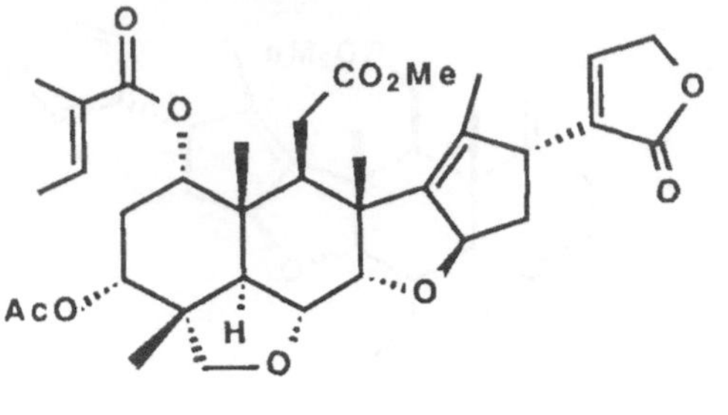

(46)

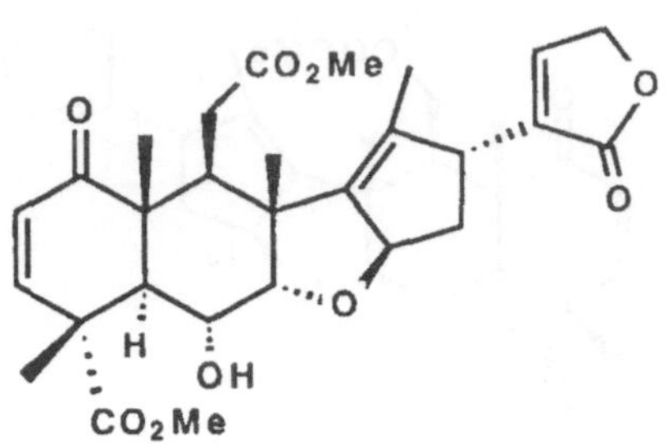

(47)

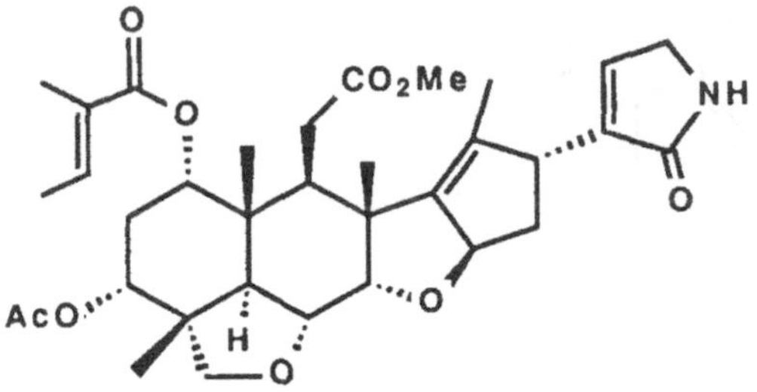

(48)   salannolactam I

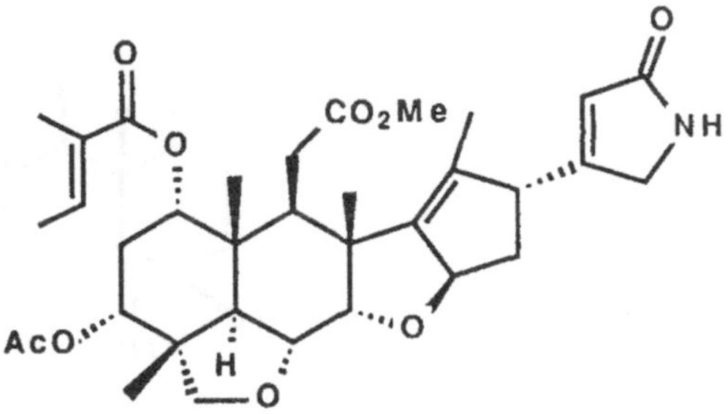

(49)   salannolactam II

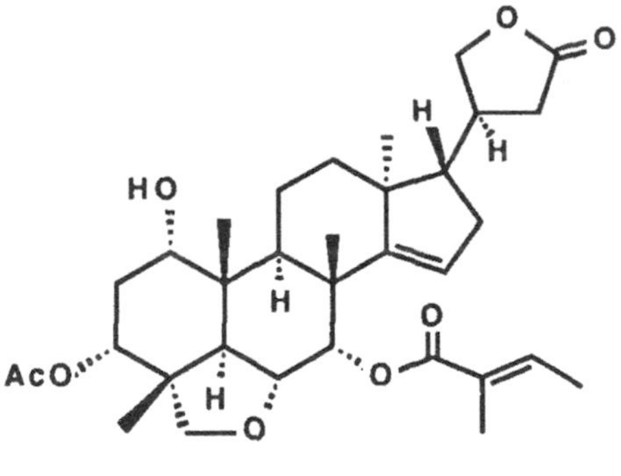

(50)   3-acetyl-7-tigloylvilasinnin lactone

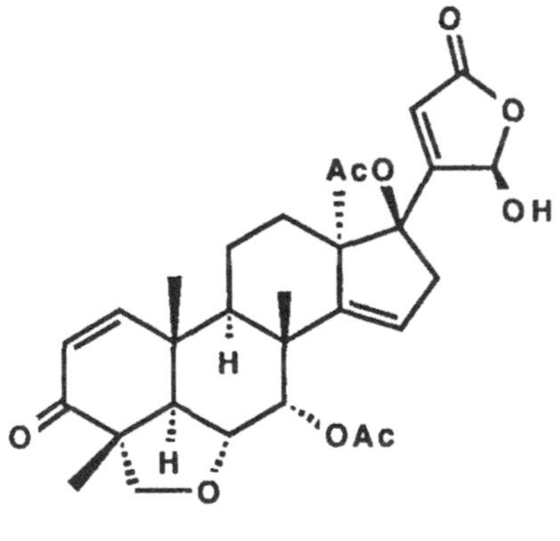

(51)   isonimolicinolide

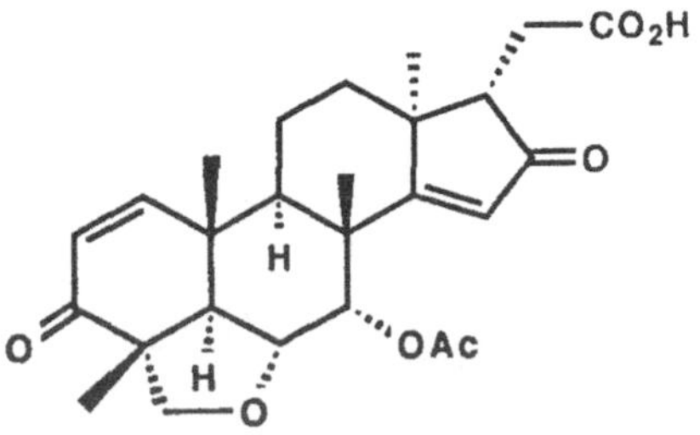

(52)   nimolicinoic acid

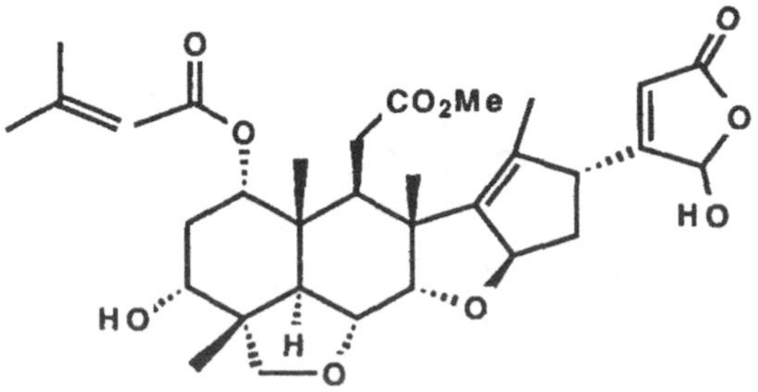

(53)   isoazadirolide

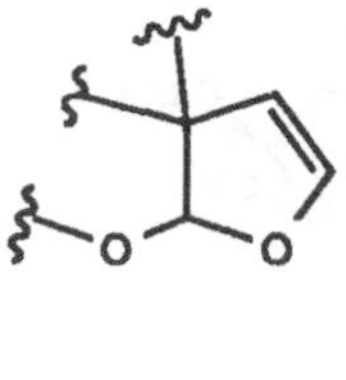

(54)

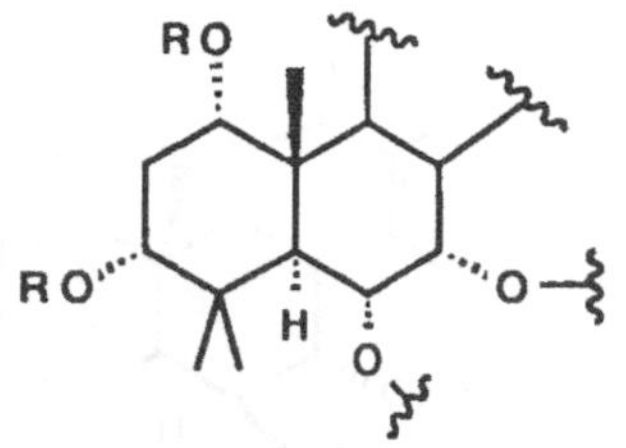

(55)

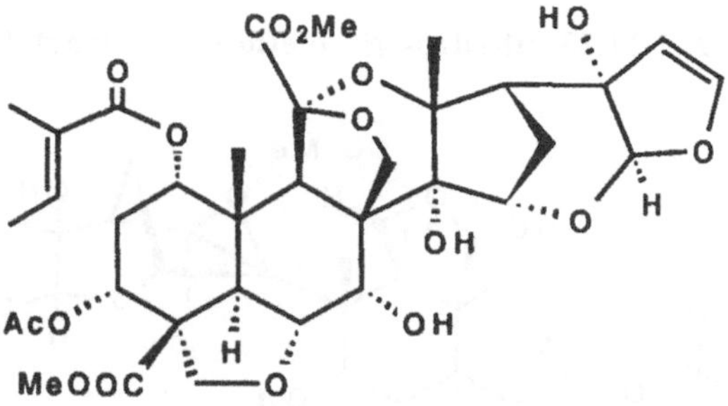

(56) Nakanishi's azadirachtin

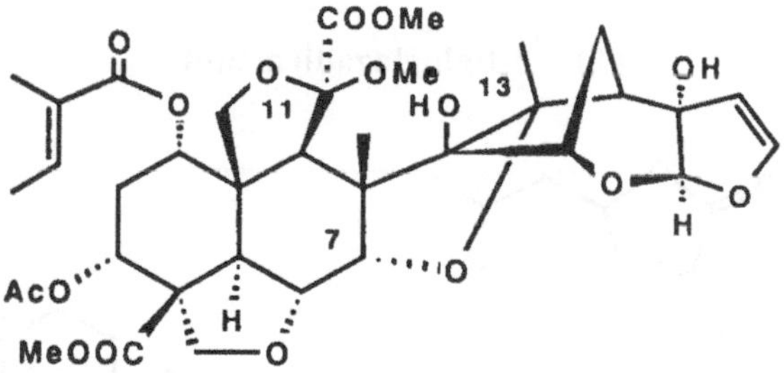

(57)   1-tigloyl-3-acetyl-11-methoxyazadirachtinin

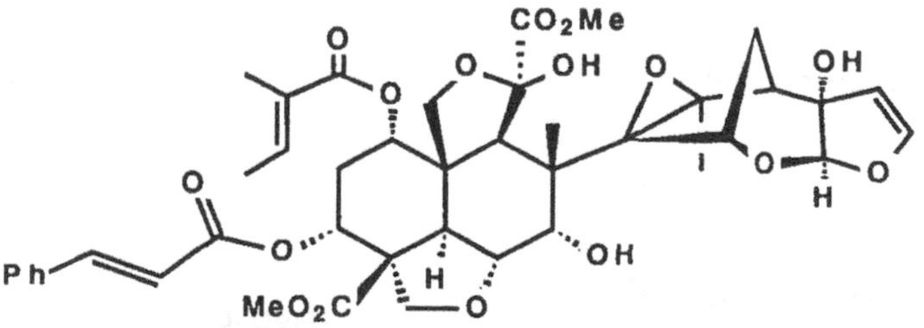

(58)   3-deacetyl-3-cinnamoylazadirachtin

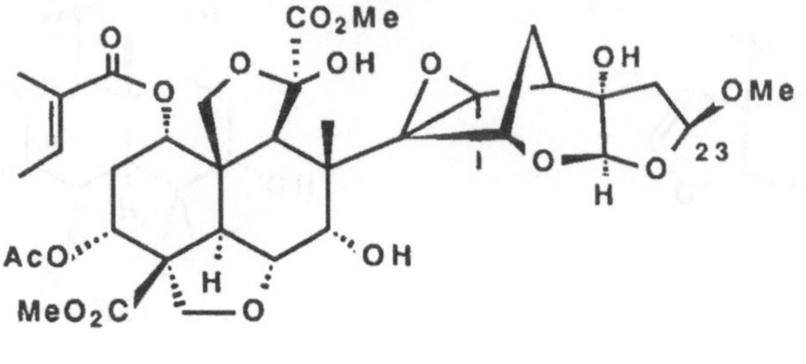

(59)   22,23-dihydro- β -methoxyazadirachtin

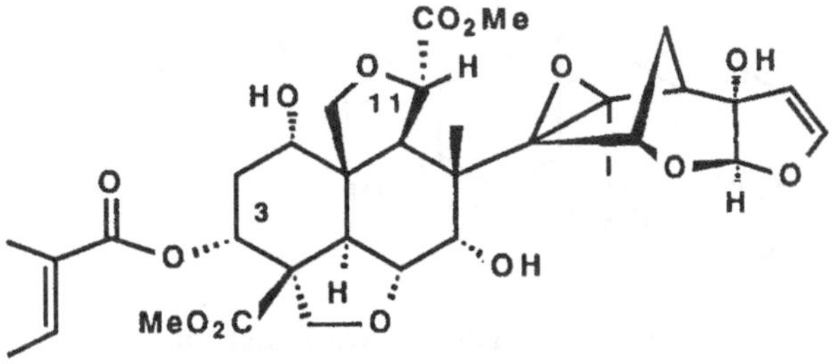

(60)   3-tigloylazadirachtol

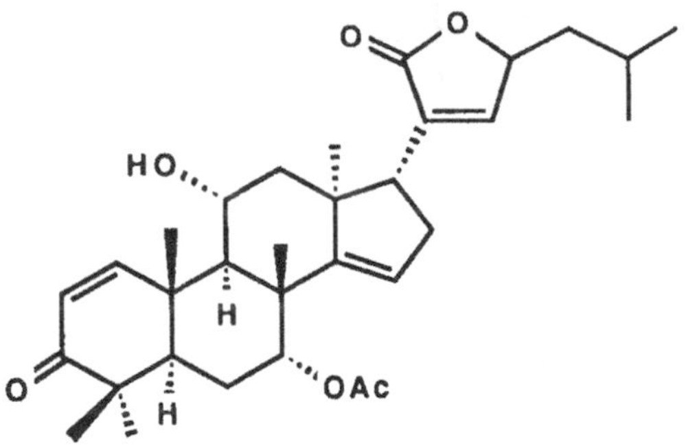

(61)   azadirachtol

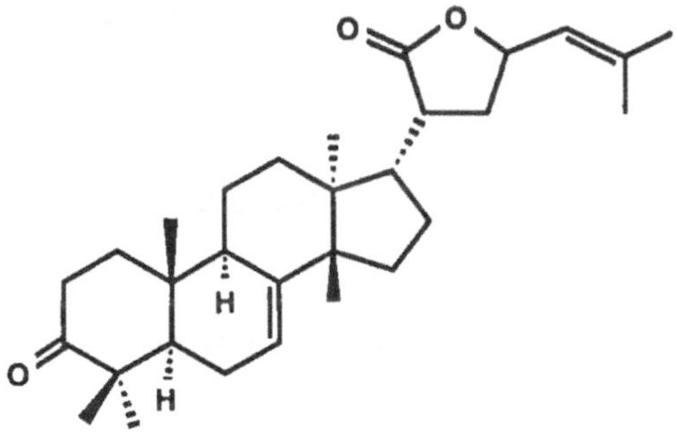

(62)   nimolinone

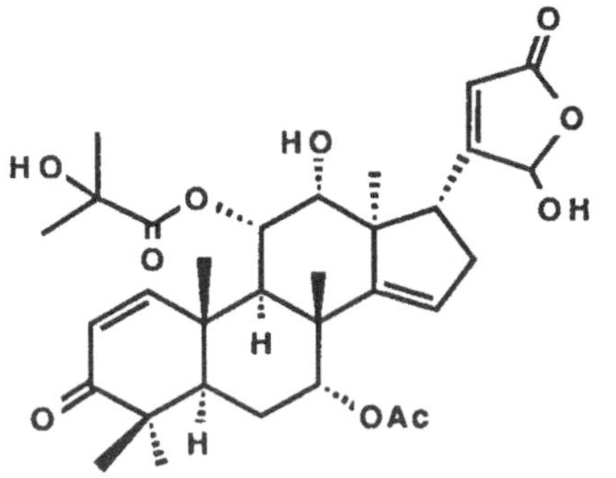

(63)   isonimbocinolide

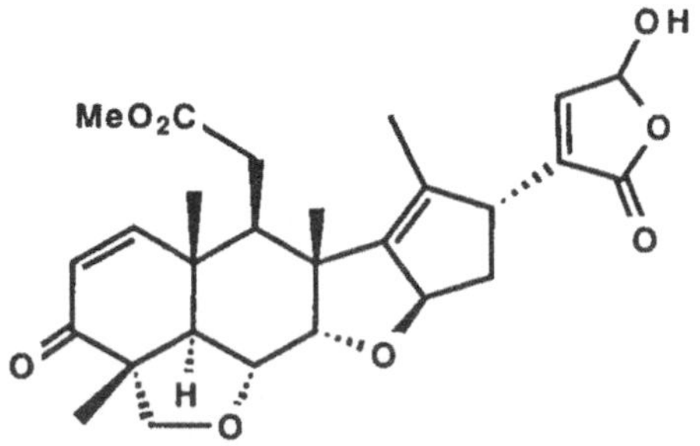

(64)   margosinolide

REFERENCES

1. **Butterworth, J. H. and Morgan, E. D.,** Isolation of a substance that suppresses feeding in locusts, *J. Chem. Soc., Chem. Commun.,* 23, 1968.
2. **Skellon, J. H., Thornburn, S., Spence, J., and Chatterjee, S. N.,** The fatty acids of neem oil and their reduction products, *J. Sci. Food. Agric.,* 13, 639, 1962, and references cited therein.
3. **Sengupta, P., Choudhiri, S. N., and Khastigir, H. N.,** Terpenoids and related compounds-I, *Tetrahedron,* 10, 45, 1960.
4. **Siddiqui, S.,** A note on the isolation of three new bitter principles from the neem oil, *Curr. Sci.,* 11, 278, 1942.
5. **Bloch, E., Ahmad, S., Catalfamo, J. L., Jain, M. K., and Apitz-Castro, R.,** Antithrombic organosulfur compounds from garlic: structural, mechanistic and synthetic studies. *J. Am. Chem. Soc.,* 108, 7045, 1986.
6. **Arigoni, D., Jeger, O., and Ruzicka, L.,** On triterpenes: the structure and configuration of tirucallol, euphorbol and elemadieneolic acid, *Helv. Chim. Acta,* 38, 222, 1955.
7. **Lavie, D., Jain, M. K., and Shpan-Gabrielith, S. R.,** A locust phagorepellent from two *Melia* species, *J. Chem. Soc., Chem. Commun.,* 910, 1967.
8. **Siddiqui, S., Mahmood, T., Siddiqui, B. S., and Faizi, S.,** Isolation of a triterpenoid from *Azadirachta indica, Phytochemistry,* 25, 2183, 1986.
9. **Buchanan, J. G. St. C. and Halsall, T. G.,** The synthesis of possible intermediates in the biogenesis of tetranortriterpenes, *J. Chem. Soc., Chem. Commun.,* 48, 1969.
10. **Buchanan, J. G. St. C. and Halsall, T. G.,** The synthesis of the simplest Meliacins (limonoids) from tetranortirucallene, *J. Chem. Soc., Chem. Commun.,* 242, 1969.
11a. **Lavie, D. and Jain, M. K.,** Tetranortriterpenoids from *Melia azadirachta* L., *J. Chem. Soc., Chem. Commun.,* 278, 1967.
11b. **Lavie, D., Levy, E. C., and Jain, M. K.,** Limonoids of biogenetic interest from *Melia azadirachta* L., *Tetrahedron,* 27, 3927, 1971.
12. **Kraus, W. and Cramer, R.,** 17-Epiazadiradione and 17b-hydroxyazadiradione, two new products from *Azadirachta indica* A. Juss, *Tetrahedron Lett.,* 2395, 1978.
13. **Connolly, J. D., Handa, K. L., and McCrindle, R.,** Further constituents of nim oil: the constitution of meldenin, *Tetrahedron Lett.,* 437, 1968.
14. **Narayanan, C. R., Pachapurkar, R. V., and Sawant, B. M.,** Nimbinin: a new tetranortriterpenoid, *Tetrahedron Lett.,* 3563, 1967.
15. **Kraus, W., Cramer, R., and Sawitski, G.,** Tetranortriterpenoids from the seeds of *Aza-* *dirachta indica, Phytochemistry,* 20, 117, 1981.
16. **Pachapurkar, R. V., Kornule, P. M., and Narayanan, C. R.,** A new hexacyclic tetranortriterpenoid, *Chem. Lett.,* 357, 1974.
17. **Kraus, W. and Cramer, R.,** New tetranortriterpenoids with insect antifeedant activity from neem oil, *Liebigs Ann. Chem.,* 181, 1981.
18. **Narayanan, C. R., Pachapurkar, R. V., Sawant, B. N., and Wadia, M. S.,** Vepenin, a new constituent of neem oil, *Indian J. Chem.,* 7, 187, 1969.
19. **Podder, G. and Mahoto, S. B.,** Azadirachtanin A. New liminoid from the leaves of *Azadirachta indica, Heterocycles,* 23, 2321, 1985.
20. **Sengupta, P., Sengupta, S. K., and Khastigir, H. N.,** Terpenoids and related compounds. II. Investigations on structure of nimbin, *Tetrahedron,* 11, 67, 1960.
21. **Narasimhan, N. C.,** The functional groups of nimbin, *Chem. Ind.,* 661, 1957.
22. **Mitra, C.,** Constituents of nim (*Melia indica*). I. Characterization of nimbin, *J. Sci. Ind. Res. India,* 15B, 425, 1956.
23. **Narayanan, C. R., Pachapurkar, R. V., Pradhan, S. K., Shah, V. R., and Narasimhan, N. S.,** Structure of nimbin, *Chem. Ind.,* 322, 1964.
24. **Narayanan, C. R., Pachapurkar, R. V., Pradhan, S. K., Shah, V. R., and Narasimhan, N. S.,** Stereochemistry of nimbin, *Chem. Ind.,* 324, 1964.
25. **Harris, H., Henderson, R., McCrindle, R. Overton, K. H., and Turner, D. W.,** Tetranortriterpenoids. VIII. The constitution and stereochemistry of nimbin, *Tetrahedron,* 24, 1517, 1968.
26. **Narayanan, C. R. and Pachapurkar, R. V.,** The structure of nimbinic acid, *Tetrahedron Lett.,* 553, 1966.
27. **Narayanan, C. R. and Pachapurkar, R. V.,** Ring D in nimbin, *Tetrahedron Lett.,* 4333, 1965.
28. **Henderson, C. R., McCrindle, R., Malera, A., and Overton, K. H.,** Tetranortriterpenoids. IX. The constitution and stereochemistry of salannin, *Tetrahedron,* 24, 1525, 1968.
29. **Ekong, D. E. U., Fakunle, C. O., Fasina, A. K., and Okogun, J. I.,** The meliacins (limonoids). Nimbolin A and B, the new meliacin cinnamates from *Azadirachta indica* L. and *Melia azedarach* L., *J. Chem. Soc., Chem. Commun.,* 1166, 1969.
30. **Ekong, D. E. U.,** Chemistry of the meliacins (limonoids). The structure of nimbolide, a new meliacin from *Azadirachta indica, J. Chem. Soc., Chem. Commun.,* 808, 1967.
31. **Kraus, W. and Cramer, R.,** Pentanortriter-

penoids from *Azadirachtia indica* A. Juss (Meliaceae), *Chem. Ber.*, 114, 2375, 1981.

32. **Bruhn, A., Bokel, M., and Kraus, W.**, 4a,6a-Dihydroxy-A-homoazadirone, a new tetranortriterpenoid from *Azadirachta indica* A. Juss (Meliaceae), *Tetrahedron Lett.*, 25, 3691, 1984.

33. **Siddiqui, S., Faizi, S., and Siddiqui, B. S.**, Studies on the chemical constituents of *Azadirachta indica* A. Juss (Meliaceae). I. Isolation and structure of a new tetranortriterpenoid nimolicinol, *Heterocycles*, 22, 295, 1984.

34. **Kraus, W., Klenk, A., Bokel, M., and Vogler, B.**, Tetranortriterpenoid-lactams with insect feeding deterrent effect from *Azadirachta indica* A. Juss (Meliaceae), *Liebigs Ann. Chem.*, 337, 1987.

35. **Bilton, J. N., Broughton, H. B., Jones, P. S., Ley, S. V., Lidert, Z., Morgan, E. D., Rzepa, H. S., Sheppard, R. N., Slawin, A. M. Z., and Williams, D. J.**, An X-ray crystallographic, mass spectroscopic and NMR study of the limonoid insect antifeedant azadirachtin and related derivatives, *Tetrahedron*, 43, 2805, 1987.

36. **Siddiqui, S., Mahmood, T., Faizi, S., and Siddiqui, B. S.**, Studies in the chemical constituents of *Azadirachta indica* A. Juss (Meliaceae). X. Isolation and structure elucidation of isonimolicinolide, the first 17-acetoxy tetranortriterpenoid, and nimolicinoic and, the first hexanortriterpenoid with an apoeuphane (apotirucallane) skeleton, *J. Chem. Soc., Perkin Trans. 1*, 1429, 1987.

37. **Siddiqui, S., Siddiqui, B. S., Faizi, S., and Mahmood, T.**, Isozadirolide, a new tetranortriterpenoid from *Azadirachta indica* A. Juss (Meliaceae); *Heterocycles*, 24, 3163, 1986.

38. **Butterworth, J. H. and Morgan, E. D.**, Investigation of the locust feeding inhibition of the seeds of the neem tree, *Azadirachta indica*, *J. Insect Physiol.*, 17, 969, 1971.

39. **Butterworth, J. H., Morgan, E. D., and Percy, G. R.**, The structure of azadirachtin; the functional groups, *J. Chem. Soc., Perkin Trans. 1*, 2445, 1972.

40. **Zanno, P. R., Miura, I., Nakanishi, K., and Elder, D. L.**, Structure of the insect phagorepellant azadirachtin. Application of PRFT/CWD carbon-13 nuclear magnetic resonance, *J. Am. Chem. Soc.*, 97, 1975, 1975.

41. **Bilton, J. N., Broughton, H. B., Ley, S. V., Lidert, Z., Morgan, E. D., Rzepa, H. S., and Sheppard, R. N.**, Structure reappraisal of the limonoid insect antifeedant azadirachtin, *J. Chem. Soc., Chem. Commun.*, 968, 1985.

42. **Broughton, H. B., Ley, S. V., Slawin, A. M. Z., Williams, D. J., and Morgan, E. D.**, X-ray crystallographic structure deter-mination of detigloyl-dihydroazadirachtin and reassignment of the structure of the limonoid insect antifeedant azadirachtin, *J. Chem. Soc., Chem. Commun.*, 46, 1986.

43. **Kraus, W., Bokel, M., Klenk, A., and Pöhnl, H.**, The structure of azadirachtin and 22,23-dihydro-23β-methoxyazadirachtin, *Tetrahedron Lett.*, 26, 6435, 1985.

44. **Kraus, W., Bokel, M., Bruhn, A., Cramer, R., Klaiber, I., Klenk, A., Nagl, G., Pöhnl, H., Sadlo, H., and Vogler, B.**, Structure determination by NMR of azadidrachtin and related compounds from *Azadirachta indica* A. Juss (Meliaceae), *Tetrahedron*, 43, 2817, 1987.

45. **Turner, C. J., Tempesta, M. S., Taylor, R. B., Zagorski, M. G., Termini, J. S., Schroeder, D. R., and Nakanishi, K.**, An NMR spectroscopic study of azadirachtin and its trimethyl ether, *Tetrahedron*, 43, 2789, 1987.

46. **Kubo, I., Matsumoto, A., and Matsumoto, T.**, New insect ecdysis inhibitory limonoid deacetylazadirachtinol isolated from *Azadirachta indica* (Meliaceae) oil, *Tetrahedron*, 42, 489, 1986.

47. **Schroeder, D. R. and Nakanishi, K.**, A simplified isolation procedure for azadirachtin, *J. Nat. Products*, 50, 241, 1987.

48. **Yamasaki, R. B., Klocke, J. A., Lee, S. M., Stone, G. A., and Darlington, M. V.**, Isolation and purification of azadirachtin from neem (*Azadirachta indica*) seeds using flash chromatography and high performance liquid chromatography, *J. Chromatogr.*, 356, 220, 1986.

49. **Warthen, J. D., Stokes, J. B., Jacobson, M., and Kozempel, M. F.**, Estimation of azadirachtin content in neem extract and formulations, *J. Liq. Chromatogr.*, 7, 591, 1984.

50. **Taylor, D. A. H.**, The chemistry of the limonoids from Meliaceae, *Fortschr. Chem. Org. Naturst.*, 45, 1, 1984.

51. **Ley, S. V. and Santafianos, D.**, unpublished observations.

52. **Yamasaki, R. B. and Klocke, J. A.**, Structure-bioactivity relationships of azadirachtin, a potent insect control agent, *J. Agric. Food. Chem.*, 35, 467, 1987.

53. **Ley, S. V.**, manuscript in preparation.

54. **Ley, S. V., Santafianos, D., Blaney, W. M., and Simmonds, M. S. J.**, Synthesis of a hydroxydihydrofuranacetal related to azadirachtin: a potent insect antifeedant, *Tetrahedron Lett.*, 28, 221, 1987.

55. **Pflieger, D., Muckensturm, B., Robert, P. C., Simonis, M. T., and Kienlen, J. C.**, Synthesis of the furo-pyran moiety of azadirachtin, *Tetrahedron Lett.*, 28, 1519, 1987.

56. **Abad, A., Brasca, M. G., Craig, D., Ley, S. V., and Toogood, P.**, submitted.

57. **Siddiqui, S. and Fuchs, S.,** Structures of new natural products from *Melia azadirachta* Linn: 17-hydroxyazadiradione, *Tetrahedron Lett.,* 611, 1978.

58. **Mitra, C. R., Garg, H. S., and Pandey, G. N.,** Identification of nimbidic acid and nimbidinin from *Azadirachta indica, Phytochemistry,* 10, 857, 1971.

59. **Mitra, C. R., Garg, H. S., and Pandey, G. N.,** Constituents of *Melia indica.* II. Nimbidic acid and nimbidinin, *Tetrahedron Lett.,* 2761, 1970.

60. **Narayanan, C. R. and Iyer, K. N.,** Isolation and characterisation of desacetylnimbin, *Indian J. Chem.,* 5, 2183, 1986.

61. **Connolly, J. D., Labbe, C., Rycroft, D. S., and Taylor, D. A. H.,** Tetranortriterpenoids and related compounds. XXII. New apotirucallol derivatives and tetranortriterpenoids from the wood and seeds of *Chisocheton paniculatus* (Meliaceae), *J. Chem. Soc., Perkin Trans. 1,* 2959, 1979.

62. **Siddiqui, S., Siddiqui, B. S., and Faizi, S.,** Studies in the chemical constituents of *Azadirachta indica.* II. Isolation and structure of the new triterpenoid azadirachtol, *Planta Med.,* 478, 1985.

63. **Siddiqui, S., Siddiqui, B. S., Faizi, S., and Mahmood, T.,** Studies on the chemical constituents of *Azadirachta indica* A. Juss (Meliceae). IV., *J. Chem. Soc. Pakistan,* 8, 341, 1986.

64. **Siddiqui, S., Faizi, S., and Siddiqui, B. S.,** Chemical constituents of *Azadirachta indica* A. Juss (Meliaceae). VII., *Z. Naturforsch.,* 41B, 922, 1986.

65. **Siddiqui, S., Faizi, S., Mahmood, T., and Siddiqui, B. S.,** Isolation of a new tetranortriterpenoid from *Azadirachta indica* A. Juss (Meliaceae), *Heterocycles,* 24, 1319, 1986.

66. **Siddiqui, S., Mahmood, T., Siddiqui, B. S., and Faizi, S.,** Two new tetranortriterpenoids from *Azadirachta indica, J. Nat. Products* 49, 1068, 1986.

67. **Grimmer, W. and Kraus, W.,** Photooxidation of 3-substituted furans in aprotic solvents, *Liebigs Ann. Chem.,* 1571, 1979.

68. **Siddiqui, S., Faizi, S., Mahmood, T., and Siddiqui, B. S.,** Margosinolide and isomargosinolide, two new tetranortriterpenoids from *Azadirachta indica* A. Juss (Meliaceae), *Tetrahedron,* 42, 4849, 1986.

69. **Lee, S. M., Balandrin, M. F., and Klocke, J. A.,** Biologically active volatile organosulfur compounds from neem seeds, *Abstr. 194th Natl. Meet., Am. Chem. Soc., Agrochem. Div.* No. 99, 1987.

70. **Devakumar, C. and Mukerjee, S. K.,** 4-Epinimbin, a new meliacin from *Azadirachta indica* A. Juss, *Indian J. Chem.,* 24B, 1105, 1985.

71. **Siddiqui, S., Siddiqui, B. S., Faizi, S., and Mahmood, T.,** Isolation of a tetranortriterpenoid from *Azadirachta indica, Phytochemistry,* 23, 2899, 1984.

3

Isomeric Azadirachtins and Their Mode of Action

Heinz Rembold
Max-Planck-Institute for Biochemistry
Martinsried, West Germany

TABLE OF CONTENTS

I. INTRODUCTION

The history of the azadirachtins is closely connected with the type of bioassay which was used for detecting these compounds. This gives a basis for understanding — or sometimes mutually misunderstanding — what the chemist, being interested in structures, or *vice versa* the biologist, being interested in their activity, understands when using the same term "azadirachtin". When Butterworth and Morgan[1] reported on the isolation of a substance from neem seeds which they named azadirachtin, they had used its feeding inhibitory effect in the desert locust, *Schistocerca gregaria*, for its purification. Tests under their standard conditions with the pure compound gave complete inhibition of the feeding response at the remarkably low concentration of 5 mg/l. Consequently, Zanno et al.[2] reported on the structure of an insect phagorepellent when they gave the first complete structure for azadirachtin.

The second act of the azadirachtin play started in 1979 when Schmutterer and Rembold[3] initiated cooperation on neem compounds which did not deter the Mexican bean beetle, *Epilachna varivestis*, from feeding and which were highly active insect growth inhibitors exclusively. Besides its more or less deterring activity, azadirachtin was also known to have growth inhibitory effects in most if not all economically important orders.[4-12] Were there also substances present in the neem seed with only one, the growth inhibitory character? Such compounds would be interesting as they would not deter the feeding insect and therefore could go unnoticed and then unfold their whole growth-inhibiting capacity in low concentrations! The authors indeed found a series of pure compounds which were contained in the polar, methanol-soluble extract from neem seed and which were not phagorepellents but which disturbed metamorphosis of *Epilachna* in minute amounts. Most of these induced in the test larvae formation of reddish-brown spots in the dorso-lateral zones of the thorax. These treated larvae did not die until about 3 weeks after appearance of such spots, most of them without further metamorphosis, as larvae or as larval-pupal intermediates. Such an effect indicated an interference of these compounds with the hormonal system.[3] The neem seed fractions also disrupted growth in other insects, again without inducing any feeding inhibition.[13] Such compounds are of interest for the chemist from the standpoint of his search for new chemical structures aiming at alternative plant protection strategies. The target of these chemicals would be the insect hormone system directly, whereas antifeedants can only indirectly cause developmental deviants,[14] their target being some chemoreceptors which then initiate hormonal effects through subsequent starvation.

The third act begins with the surprising result that the main *insect growth inhibitor* is identical with azadirachtin, a compound which until then had been treated as a *phagorepellent*.[15] A short time later it became clear that some of our unknown growth inhibitors were isomeric with the compound already known as "azadirachtin".[16,17] As will be discussed later, there is also an antifeedant effect of azadirachtin with *Epilachna* and other holometabolous insects. The required concentrations, however, considerably exceed those which induce morphogenetic disturbances. Which are the other growth inhibiting neem compounds and what is their mode of action? These points will be treated in the following sections and will introduce the first idea about structure-activity relationships in the class of naturally occurring azadirachtins. However, we shall also learn that we still have to wait for the fourth act. This will deal with the molecular target in the insect which, as a consequence of azadirachtin fed in trace amounts, completely loses control of its morphogenetic program. Nothing is known about this scenario at the moment, however.

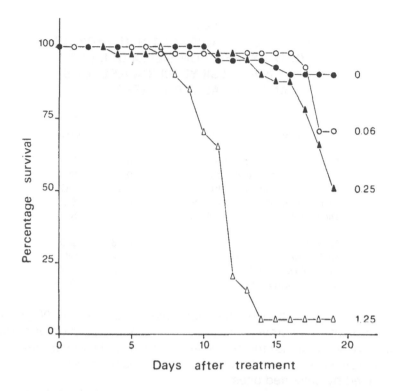

FIGURE 1. Effect of azadirachtin A in the *Epilachna varivestis* Petri dish bioassay.[13] Concentrations of the methanolic azadirachtin solution are given in parts per million. Each concentration was tested with 20 larvae and the number of survivors plotted daily, expressed as percentage.

II. STRUCTURES OF THE AZADIRACHTIN GROUP

A. THE *EPILACHNA* BIOASSAY

The azadirachtins belong to the group of C_{26}-terpenoids which exhibit a variety of biological activities. Some of them act as antineoplastic and cytotoxic agents, others as pesticides and antifeedants.[18] With these many possibilities in mind, one must carefully select a bioassay for the azadirachtins which combines high sensitivity for growth disruption with high tolerance for antifeedants. The Mexican bean beetle, *E. varivestis*, combined these two attributes under simple test conditions. Two tests have been described for routine assays: a petri dish test for individual larvae and a cage test for groups of larvae.[13] The test insects are reared on bean leaves, *Phaseolus vulgaris*, and maintained at 25°C as described by Steets.[8]

For the *petri dish test*, one bean leaf is placed in each of 20 plastic petri dish covers (9.2 cm diameter), placed upside down, containing a moist filter paper. The leaf is then covered with the same size petri dish bottom having a hole of 4-cm diameter. The test material, dissolved in 0.2 ml of methanol, is equally distributed on the exposed leaf area (12.6 cm²), and the dish is covered with another petri dish top. In each dish a weighed, freshly molted fourth instar larva is released. After 24 and 48 h, the weight gains of all larvae are calculated and the larvae collected, transferred to untreated leaves, and their development is followed. A typical test of this kind is shown in Figure 1.

For routine estimations, the *cage test* requires less expenditure of work and also eliminates a possible fumigant action of volatile by-products. An aluminum-framed cage (30 × 25 × 32 cm) with a wooden bottom, a glass top, and the sides covered with a

Table 1
R$_F$ VALUES OF AZADIRACHTIN IN VARIOUS
SOLVENT SYSTEMS ON SILICA GEL (SiO$_2$)
AND OCTADECYLSILYL-SILICA GEL (ODS)
ANALYTICAL TLC PLATES[25]

Solvent system (v/v)	R$_F$	TLC plate
Diethyl ether-methanol (99:1)	0.24	SiO$_2$
Dichloromethane-acetone (4:1)	0.30	SiO$_2$
Diethyl ether-methanol (49:1)	0.37	SiO$_2$
Dichloromethane-methanol (19:1)	0.52	SiO$_2$
Diethyl ether-methanol-acetic acid (95:5:1)	0.55	SiO$_2$
Isopropanol-*n*-hexane (11:9)	0.56	SiO$_2$
Diethyl ether-acetone (2:1)	0.74	SiO$_2$
Acetonitrile-water (9:11)	0.32	ODS
Methanol-water (3:2)	0.37	ODS

mesh-net for ventilation can be used for each assay. Together with 20 primary leaves, 10 young bean plants are sprayed with 4 ml of methanolic solution and placed inside the cage after drying. Then, 20 freshly molted fourth instar larvae are released in each cage and new grown leaves are plucked off. The treated bean plants are removed after 48 h and replaced by untreated ones.

In both petri dish and cage test, two control treatments are kept. In one control, the leaves are treated with methanol and in the other they are left untreated. The undisturbed development of the control from beginning of the fourth instar to the newly hatched adult takes 8 d. The tests have to be repeated if the concentration of the applied compound does not result in about 50% survival. On this basis, the MC$_{50}$ (50% metamorphosis inhibiting concentration) values can be calculated.

B. ISOLATION OF THE AZADIRACHTINS

Azadirachtin is difficult to isolate and the yields are usually low. Several techniques have been described for its preparation from neem seed kernels,[4,12,13,16,19-26] all of which, in principle, make use of the same techniques: solvent extraction, adsorption chromatography on silica gel, and reversed phase high pressure liquid chromatography. A typical isolation procedure[20] shall be described in the following.

Air-dried neem seed kernels (57 kg) were machine-freed from their husks (30 kg), which contained no azadirachtin. The neem kernels (27 kg) were ground in a grist mill and the powder divided into five portions. Each charge was twice extracted by stirring (precautions such as the use of an air-driven stirrer, grounding of the solvent, and continuous removal of evaporating hexane must be carefully observed) with 10 l of hexane, filtered, and the hexane extracts combined (12.0 kg neem oil). The dry, defatted neem kernel powder (15 kg), divided into three portions, was then twice extracted with 10 l each of acetone, using the same equipment. After evaporation of the acetone, 1.67 kg of residue was obtained, which contained the entire azadirachtin fraction as checked by thin layer chromatography (TLC, solvent: chloroform/acetone, 7:3; for other solvent systems see Table 1) and high performance liquid chromatography (HPLC).

For the first chromatographic step, the acetone-soluble material (1.67 kg) was dissolved in methanol, 1.5 kg of silica gel was added, and the solvent was removed *in vacuo* with a rotary evaporator. One fifth of the powder was layered on top of a silica

Table 2
PURE AZADIRACHTINS AS
ISOLATED FROM 27 KG
NEEM SEED[20]

Isomer	Amount (mg)
A	3500 mg
B	700
C	4.7
D	3.8
E	9.4
F	4.5
G	3.1

gel column (30 cm length, 10 cm diameter) which was equilibrated with petroleum ether/ethyl acetate (7:3). For elution of the first fraction, 4 l of the same solvent were passed through the column (yield 1.14 kg). The second fraction was eluted with 4 l of petroleum ether/ethyl acetate (1:4, yield 0.17 kg). Only the second fraction contained the azadirachtins. For removal of some apolar material, this fraction was distributed in two portions between the two phases of a methanol/water/hexane (4:1:5) mixture. About 95% of the azadirachtins were in the methanol phase (88 g), whereas 79 g of biologically inactive material was contained in the hexane fraction.

Of the active material, 7 g each was further purified by reverse phase chromatography on a 100-g SiO_2-RP8 (Lichroprep®, Merck) column (300 mm length, 25 mm diameter), and isocratically eluted with a mixture of methanol/water (7:3). Each 15-ml fraction was controlled by TLC and the azadirachtin-containing fractions were pooled. From the total starting material of 88 g, 15.5 g of an azadirachtin-containing fraction was obtained. Azadirachtin A was the major active compound and azadirachtin B the minor active compound, as checked by the *Epilachna* bioassay. The two compounds were isolated in a proportion of about 5:1.

Azadirachtins A and B[17] were separated by preparative HPLC on a SiO_2-RP8 column (540 mm length, 54 mm diameter) with methanol/water (43:57) as isocratic solvent.[27] For each chromatography, about 5 g of the azadirachtin fraction was applied. Each fraction corresponded to 500 ml of eluent which was collected within 9 min. The maximum azadirachtin A was eluted after 284 min and that of azadirachtin B after 397 min. Each fraction was controlled by analytical HPLC and side fractions were rechromatographed. Final yield was 3.50 g azadirachtin A and 0.70 g azadirachtin B. Of the biologically active side fractions, the five azadirachtins C to G were isolated by use of preparative TLC and semipreparative HPLC, followed by their growth inhibiting activity in the *Epilachna* bioassay. All these growth inhibiting azadirachtins are present in the acetone-soluble fraction in minute amounts only (Table 2).

C. STRUCTURE OF THE AZADIRACHTINS

Azadirachtin A (Structure I) is the predominant growth-inhibiting neem compound. Its former structure as proposed by Zanno et al.[2] has recently been reassigned by three laboratories[21,28,29] and now unequivocally gives the basis for a structural elucidation of the other isomeric azadirachtins by NMR spectroscopy. Details of their chemical structures are given in Chapter 2.

Due to their very similar chemical and physical properties, it was not until 1983 that "the azadirachtin" was determined to be a mixture of several isomers. They all share a very similar chemical structure and biological activity and therefore can be treated as

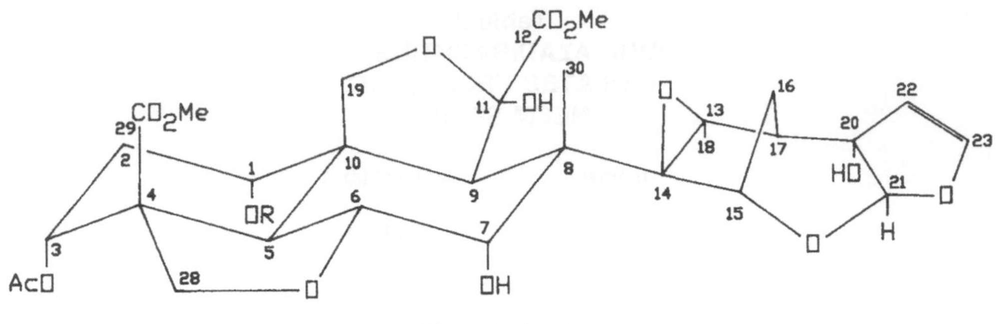

R = tigloyl

STRUCTURE I

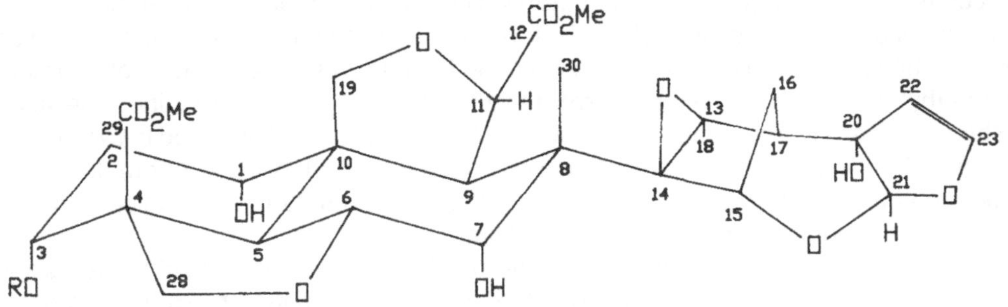

R = tigloyl

STRUCTURE II

a group of azadirachtin isomers, the predominant ones being azadirachtins A and B.[16,17] Until now, at least five have been purified and, with the exception of one (azadirachtin C), a structure based on NMR data can be proposed. In addition, by cautious chemical modification of the mother compounds, 9 more isomers were prepared, and these 16 highly active insect growth inhibitors form the basis for a first minimal structure which may open a synthetic approach for simpler structures with the same biological activity. The structures of these azadirachtins shall be discussed briefly.

Azadirachtin B (Structure II) was isolated by following its growth inhibitory activity in the *Epilachna* bioassay[17,27] and, without data of their biological activity, also in two other laboratories. Klenk et al.[30] reported on the structure of 3-tigloylazadirachtol and Kubo et al.[31] reported on deacetylazadirachtinol. According to their NMR spectra, the three compounds are identical. There are three structural differences if azadirachtins A and B are compared with one another. In azadirachtin B, the tigloyl side chain is located at position 3, the molecule has a free 1-hydroxy group and position 11 is reduced to the deoxy compound. Recently, a similar molecule has been described (3-deacetyl-11-desoxyazadirachtin),[28] however, still having the tigloyl side chain in position 1 and a free 3-hydroxyl group; it has also been isolated from neem seed.

For *azadirachtin C*, only a partial structure can be given, with the *trans*-decalin ring substituted as in azadirachtin A.[20]

Azadirachtin D (Structure III, 1.c.[32]) differs from azadirachtin A by reduction of the ester group in position 4 to a methyl group. All the rest of the molecule is unchanged if compared with the main compound of this series. A similar compound (1-cinnamoyl-3-feruloyl-11-hydroxymeliacarpin) which differs from azadirachtin D only in the substi-

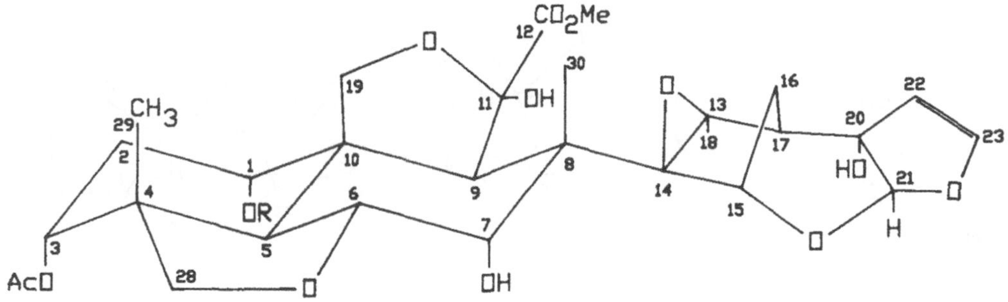

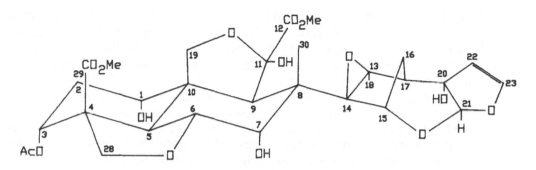

R = tigloyl

STRUCTURE III

STRUCTURE IV

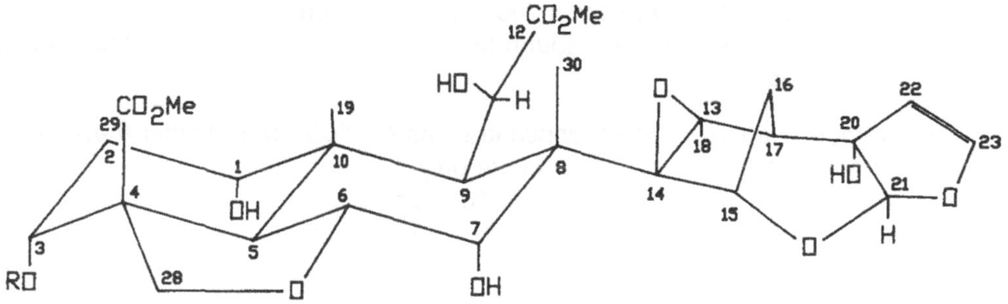

R = tigloyl

STRUCTURE V

tuents at positions 1 and 3, has been isolated from *Melia azedarach* leaves.[33] *Azadirachtin E* (Structure IV) is the naturally occurring detigloylazadirachtin A.

Natural derivatives of azadirachtin B are the isomers F and G. In *azadirachtin F* (Structure V), the ether bridge in position 19 is reduced and opened by formation of a C-19 methyl group. *Azadirachtin G* (Structure VI) is an isomer of azadirachtin B with a double bond formed instead of the epoxide ring and with a hydroxy group in position 17.

Besides the isolation of six neem seed compounds of the azadirachtin series, Forster[20] also degraded the azadirachtin A and B molecules and obtained a series of derivatives by minor modification of the respective mother compounds. 22,23-Dihydroazadirachtins

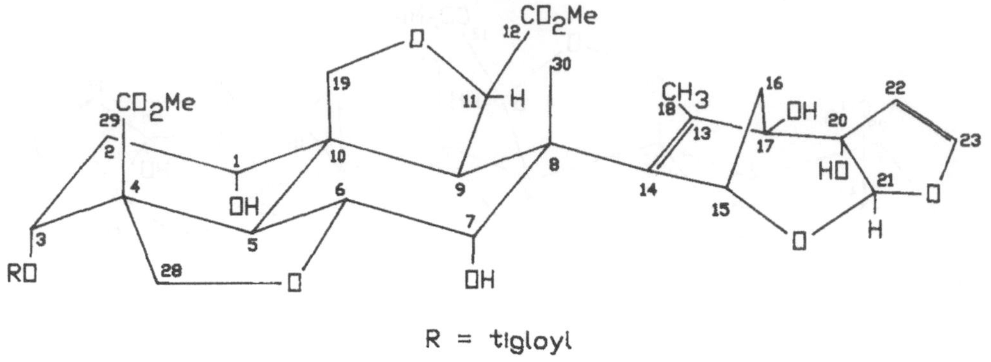

R = tigloyl

STRUCTURE VI

A and B are prepared by catalytic hydrogenation, and 11-acetylazadirachtin A by reaction of azadirachtin A with acetic anhydride.[34] 3-Deacetylazadirachtin can be prepared by saponification of the mother compound with sodium methylate and, the two isomers α- and β-23-ethoxy-22,23-dihydroazadirachtin A by reaction of azadirachtin A with acetyl chloride in chloroform. Starting from azadirachtin B, besides the 22,23-dihydro compound, 3-(2-methylbutyryl)-3-detigloyl-22,23-dihydroazadirachtin B is formed by hydrogenation with excess platinum catalyst. Detigloylazadirachtin B was prepared by hydrolysis of azadirachtin B in 10% aqueous K_2CO_3. If azadirachtin B is treated with acetyl chloride under mild conditions, α-ethoxy-22,23-dihydro-13,14-deepoxy-17-hydroxyazadirachtin is formed. Most of these reactions result only in low yields, thus demonstrating the extreme chemical lability of the azadirachtin molecule.

D. ACTIVITY IN THE *EPILACHNA* BIOASSAY

All the azadirachtins already tested induced three different effects, depending on the amount of substance applied to the fourth instar *Epilachna* larva, under the standard conditions of the *cage test*:

1. Toxic effects if applied in high concentrations (>1000 ppm). Under these conditions, the larvae survive for a few hours only.
2. In concentrations beginning from 10 to 100 ppm, the azadirachtins are very active phagodeterrents, combined with growth-disrupting activity.
3. In concentrations between 1 and 10 ppm, all the azadirachtins are growth inhibitors without any phagodeterring effect. This is exactly the concentration range where they can be described as ideal insect growth inhibitors.

The effects which are induced in the *Epilachna* larvae after feeding one of the azadirachtins in parts-per-million quantities in the cage test, are as follows:[20,35]

1. They are arrested in the larval stage without any sign of a larval-pupal molt. A few days after the azadirachtin uptake, they refuse to feed even on untreated bean plants and die of starvation.
2. Larvae which have reached the prepupal stage either dry out or form surface holes from which body fluid flows out. In both these cases they die as pharate pupae.
3. Metamorphosis to the normal pupal stage is possible. The adult beetle, however, is unable to shed the pupal integument and therefore dies.
4. An adult hatches with deformations of the wings, or the masticatory organs, or with black legs. Such animals die within a few hours.

Table 3
MC_{50} VALUES (PPM) FOR THE NATURAL ISOMERIC
AZADIRACHTINS, SOME OF THEIR CHEMICAL
DERIVATIVES, AND OF SALANNIN, AS
CALCULATED FROM *EPILACHNA* BIOASSAYS[20,35]

Azadirachtins	n	MC_{50}	sign.
aza A	40	1.66	ss
22,23-Dihydro-aza A	6	1.26	w
11-Acetyl-aza A	13	8.68	s
3-Deacetylaza A	31	0.38	ss
23α-Ethoxy-22,23-dihydro-aza A	19	0.74	s
23β-Ethoxy 22,23-dihydro-aza A	7	0.52	s
aza B	19	1.30	ss
22,23-Dihydro-aza B	27	0.28	ss
3-Detigloyl-3-(2-methylbutyryl)-22,23-dihydro-aza B	17	0.45	ss
3-Detigloyl-aza B	13	0.08	s
23α-Ethoxy-22,23-dihydro-13,14-deepoxy-17-hydroxy-aza B	8	>100	
aza C	8	12.97	s
aza D	15	1.57	ss
aza E	25	0.57	ss
aza F	18	1.15	s
aza G	16	7.69	s
salannin	4	>100	

Note: aza = azadirachtin; n = number of individual cage tests; sing. = statistical significance; SS = > 99.9%; S = 99—99.9%; w = 95—99%.

For a quantitative expression of these metamorphosis-preventing activities, all the effects described above are classed under the term "growth inhibition". The azadirachtin is applied under standard conditions in the *Epilachna* cage test at several concentrations. We name that concentration, which inhibits normal development by 50% as the "50% metamorphosis inhibiting concentration" (MC_{50}). This specific value can be calculated after Noack and Reichmuth.[36] The MC_{50} values estimated for all of the natural azadirachtins isolated by us from neem seed, as well as for the derivatives described above, are summarized in Table 3.

E. STRUCTURE-ACTIVITY RELATIONSHIPS

When comparing the 17 compounds collated in Table 3, all those products with a biological activity <10 ppm have several structural features in common. Although only 14 molecules with again only limited structural differences can be compared with each other, they all have some distinct properties in common which, if absent, abolish the growth-inhibiting activity in the *Epilachna* bioassay.

1. The type of substitution at the decalin rings A and B is important for the biological effect. Creation of a free 3-hydroxy group by deacetylation of azadirachtin A, increases the MD_{50} value about 4-fold, and hydrolysis of the 3-tigloyl moiety from azadirachtin B results in a 16-fold increase in growth inhibitory activity. There is no activity difference between azadirachtin A and its three isomers D, E, and F, with values between 1.15 and 2.80 ppm, all having the 3-hydroxy group substituted by either acetate or tiglate. Highest activity is found with both of the 1- and 3-

hydroxy groups being unsubstituted. Also, the ecdysteroid molecule has a decalin moiety and carries two hydroxy groups in the A ring. There is a fundamental difference between the ecdysteroid and the azadirachtin structures however, the latter having rings A and B *trans*-connected and the ecdysteroids *cis*-connected. However, if their molecular models are compared, the two hydroxy groups in ring A are about the same distance from each other in the azadirachtin and in the ecdysteroid molecule. This point will be discussed later once again.

2. The 22,23-double bond of the dihydrofuran ring is present in all the naturally occurring azadirachtins thus far isolated. It can easily be removed either by hydrogenation or by addition of alcohol. In the latter case, biological activity is increased in all the molecules as is apparent from Table 3. The 22,23-double bond therefore appears to be nonessential. Comparison of the azadirachtin and ecdysteroid molecular models might also indicate some similarities between the ecdysteroid side chain of eight carbon atoms and that of the four furan carbons in azadirachtin, due to the fact that elongation of this carbon skeleton by addition of an ethoxy group significantly increases biological activity.

3. Structural variation at position 11 also affects the growth inhibitory activity of the azadirachtins. Activity of azadirachtin B (1.30 ppm) and of its derivative, azadirachtin F (1.15), is significantly higher than that of azadirachtin A (1.66) and especially of its 11-acetyl derivative (8.68). It seems to be of importance for developing high biological activity that a group which takes up too much space at position 11 should not be present.

4. The structural element most critical for any growth inhibitory activity of the azadirachtins seems to be the 13,14-epoxy group, which becomes evident from a comparison of the two compounds free of the oxirane ring, i.e., 23α-ethoxy-22,23-dihydro-13,14-deepoxy-17-hydroxyazadirachtin B and salannin,[37] both of which are completely inactive as insect growth inhibitors in the *Epilachna* bioassay. Also the two azadirachtins C (12.97 ppm) and G (7.69 ppm) exhibit significantly reduced biological activity if their MC_{50} values are compared. The latter molecule has, however, as do the other 12 active azadirachtins, a ketal function at position C-21 and a free hydroxy group at C-7. Comparison of the azadirachtin G structure and its activity with 23α-ethoxy-22,23-dihydro-13,14-deepoxy-17-hydroxyazadirachtin B, which is inactive, indicates a more complicated structural relationship.

On the basis of these, admittedly preliminary, observations, a reduced structure for the biologically active portion of the azadirachtin molecule can be proposed (Structure VII).

This reduced azadirachtin structure calls special attention to the seven isomeric azadirachtins which have been isolated thus far from neem seed and the growth-inhibiting activity which has been documented. One may predict from this structure that 3-deacetyl-11-desoxyazadirachtin[28] and 1-cinnamoyl-3-feruloyl-11-hydroxymeliacarpin[33] are also active as insect growth inhibitors. On the basis of the proposed minimal structure the following statements can be made:

1. The decalin ring must be substituted at positions 1 and 3 by hydroxy groups, either one or both of which are esterified or both are unsubstituted. The two free hydroxy groups yield the highest biological activity.

2. An epoxy group must be present in correct steric distance from the two hydroxy groups. Additional effects of a hydroxy group in position 7 and/or a ketal function in position 21 are possible.

3. The reduced dihydrofuran ring or a side chain, again in correct steric distance from the epoxy group, increases the growth regulating activity of the azadirachtin analogue.

STRUCTURE VII

Although a synthetic compound of this structure does not seem to be economically feasible for practical use in plant protection, it will be helpful for further studies on even simpler and still bioactive compounds. Here, one wonders about the function of such substituents as tiglate or acetate may be for the biological activity of the azadirachtin molecule. A synergistic function in penetrating the lipophilic insect cuticle may be one reason. Another may be some sort of slow release of the bioactive compound by ester hydrolysis in the gut or only in the target organ. A third possibility may be an alkylation reaction through splitting of the epoxy ring which would explain the organ specificity and high biological activity of the azadirachtin molecule. Some of these questions will be answered in the following section on the azadirachtin mode of action. Others must remain open for further investigations.

III. MODE OF AZADIRACHTIN ACTION

It has already been discussed in the preceding sections that azadirachtin was first purified from crude neem seed extracts by following its feeding inhibitory activity in *Schistocerca gregaria*.[1] This antifeedant activity of azadirachtin was demonstrated for several other insect species.[12,38-41] However, as exemplified by the quantitative *Epilachna* biotest, this inhibitory effect is dose-dependent, diminishes with reduced concentrations, and is even absent at concentrations which still induce malformations and growth inhibition. As accentuated at the beginning, ignoring the Janus-faced character of feeding and, respectively, growth inhibitors can be the basis for misinterpreting some physiological effects of these compounds. Salannin, another compound from neem seeds, is a highly potent feeding deterrent for house flies[23] but it does not interfere with growth and development, whereas azadirachtin does both, depending on the amount of substance taken up. Azadirachtin seems to inhibit feeding at much lower concentrations in hemimetabolous than in holometabolous insects. On the other hand, even in the heteropteran *Rhodnius prolixus*, only high doses of azadirachtin A or B had an antifeedant effect when given through a blood meal, whereas molt inhibition was observed at hundred- to thousandfold lower doses.[42] There are other examples showing higher effects as a phagorepellent for azadirachtin than as a growth inhibitor. Meisner et al.[43] reported on the feeding response of *Spodoptera litura* and *Earias insulana* larvae, both of which showed a much higher inhibition with azadirachtin than with salannin.

Kubo and Klocke[44] tested the efficacy of azadirachtin as a phago- and growth inhibitor in a cotton leaf- and cotyledon-disk feeding assay. All of the four lepidopteran pest larvae (*Heliothis zea, H. virescens, Spodoptera frugiperda,* and *Pectinophora gossypiella*) were inhibited from feeding at lower azadirachtin doses than those needed for growth inhibition. Due to its systemic action, the compound protected the plant material after it was systemically carried to the cotyledon as effectively as after topical application in the leaf test. These results clearly show the antagonistic effects of azadirachtin under certain test situations, the growth inhibitory activity coming on only if the compound has been fed before and the feeding inhibitory activity which, due to the appearance of new foliage, may be effective for only a short period of time if no systemic effect is possible.

What is the growth disrupting mode of azadirachtin action? Interestingly enough, pure azadirachtin as well as other partially purified neem seed fractions, some of which belong to the azadirachtin group, were almost equally active against spinning, nonfeeding *Ephestia kuehniella* and against feeding *Apis mellifera* larvae, inducing a delayed larval development in both of them.[45] Similar results have been reported after feeding azadirachtin to larvae of the Asiatic corn borer, *Ostrinia furnacalis*.[46] This excludes growth inhibitory effects of azadirachtin by an interference with chitin synthesis or, since the spinning *Ephestia* larvae had already stopped feeding before treatment, by an indirect effect on hormonal regulation through starvation. There must be a more direct action on the regulatory system of larval morphogenesis.

A. EFFECT ON ENDOCRINE EVENTS OF THE INSECT LARVA

Very soon after its isolation was reported, azadirachtin, apart from its antifeeding effect, was also found to cause disorders in some phytophagous insects.[6,47] Treatment of the insects and/or their food with the pure compound or with azadirachtin-containing extracts caused growth inhibition, malformations, mortality, and reduced fecundity.[3,7-9,48-52] Similar morphogenetic defects could also be induced by synthetic hormone mimics and it was obviously concluded that azadirachtin could function like the ecdysoids, which were known to be present in many plants. Such a mode of action for azadirachtin seemed to be supported by the finding of Käuser and Koolman[53] that azadirachtin reduced ponasterone A-binding to ecdysteroid receptors in *Calliphora* epidermis *in vitro*. Competition studies with the pure ecdysone receptor did not show any difference and therefore seem to disprove this assumption.[54] The stereochemical differences between the two molecules, azadirachtin and ecdysone, which have been discussed already, make this negative result intelligible. In addition to the direct competition experiment, there is also no effect by azadirachtin on prothoracic gland secretion nor does it interfere with prothoracotropic hormone (PTTH) function. For their studies Koul et al.[55] used *in vitro* techniques. They measured ecdysone secretion by fifth instar *Bombyx* prothoracic glands which were incubated in the presence and absence of PTTH, azadirachtin, and mixtures of both. No marked difference was found when either prothoracic glands as controls, or activated with PTTH, were exposed to azadirachtin. These results prove beyond doubt that the compound exerts neither a direct, competitive effect on prothoracic gland secretion, nor does it block any PTTH receptor sites on this endocrine organ.

What are the effects of azadirachtin, either fed or topically applied, on the growing larva? Some of them have already been mentioned, such as dose-dependent reduction of larval development, disturbance of metamorphosis, and mortality. A detailed study of the effects on the hormone system of the last-larval instar of *Locusta migratoria* showed a typical dose-dependence on the responding animals.[56] At a dose of 0.6 μg azadirachtin/g only 10% of the animals showed a reaction whereas 2 μg/g elicited a maximal response and no larva was able to undergo or terminate ecdysis. The intermolt of azadirachtin-injected larvae varied between 8 and 60 d. That of control (fourth and fifth instar) larvae lasted for 6 and 9 d, respectively. Similar results were described by Koul et al.[55] after injection of *Bombyx mori* fifth instar larvae. Azadirachtin was extremely effective in producing pupal deformities and larval growth inhibition at 1 and 2 μg/g, respectively, when given by injection. Even more dramatic effects are induced in *Rhodnius prolixus* after the substance is taken up with a blood meal.[57] The effective dose that prevented ecdysis in 50% of the insects (ED_{50}) was 4×10^{-4} μg/ml of blood, and doses higher than 1 μg/ml inhibited ecdysis by 100%. A single treatment with azadirachtin was enough to inhibit any molt even after a period of 5 months, although the bugs survived and were fed five times on untreated blood. Dorn et al.[58] injected azadirachtin into newly molted last-instar larvae of *Oncopeltus fasciatus* and also found the induction of permanent larvae with a more than fourfold lifespan after an injection of the high dose of 8 μg/larva. Zebitz[59] found that continuous exposure of *Aedes aegypti* larvae to crude azadirachtin delayed growth and caused morphogenetic disturbances in this dipteran insect.

Such reactions as inhibition of metamorphosis indicate an interaction of azadirachtin with the hormone system of the treated larvae. There is a pronounced effect of azadirachtin on control of ecdysteroid titer as first demonstrated in fifth instar *Locusta migratoria*.[56,60] The authors explain this effect as an interference of the compound with the neuroendocrine system of the larvae. This argument is supported by histological studies which clearly show an increase of paraldehyde-fuchsin stainable material in the neurosecretory cells of the *Pars intercerebralis* of azadirachtin-treated last-instar locusts.[50,56,60] Concomitantly, also with the ecdysteroid the juvenile hormone synthesis is affected by azadirachtin.[61] These effects can be discussed under different aspects. The dose-dependent sharp increase of larval response to the compound with an ED_{50} value around 1.5 μg/g seems to be due to the altered feeding behavior that leads to high mortality, possibly by influencing neural control centers. This is consistent with observations of Redfern et al.,[48] who found persistent feeding inhibition in *Spodoptera frugiperda* larvae after transfer from an azadirachtin-treated to an untreated diet. The dose necessary for inducing such effects seems to be much lower than indicated by the ED_{50} value, as found with tracer studies.[17] [22,23-^{3}H$_2$]dihydroazadirachtin A, if injected in a dose which induces hormonal disturbances, is excreted again, within 7 h to a great extent, most of it unchanged. Such a result speaks in favor of a high-affinity azadirachtin-binding protein which, however, has not yet been found and which obviously is not identical with an ecdysteroid receptor, as already discussed. The effect on neural control centers is also indicated by changes in behavior after azadirachtin treatment. As discussed, application of the compound to early larval stages of *Locusta* extends the duration of the larval stage to several weeks. Such larvae show a sexual behavior like adults[62] and flight pattern formation, the flight muscle activity resembling the flight motor pattern of young locusts.[63] Azadirachtin treatment of *Leucophaea maderae* shortens the period length of the locomotor activity rhythm in the circadian rhythm and induces splitting of this rhythm into two components. Such effects persist for an extended period of several weeks.[64,65]

The results demonstrate that azadirachtin is extremely effective in regulating growth and behavior in *Locusta* at low doses, and that feeding inhibition is not the primary cause for growth disruption. Inhibition of ecdysis in azadirachtin-treated larvae is due to interference with the hormonal control of molting. The ecdysteroid titer in the treated larvae reveals a close relationship between endocrine conditions and morphogenetic processes. Shift or even complete disappearance of the molting-hormone titer is the result of azadirachtin application.[56] Such a shift of the ecdysone titer can affect feeding behavior. Ecdysone administration at physiological concentrations (35 μg/fifth instar

Locusta migratoria nymph) reduced the live weight gain and all the treated hoppers failed to molt.[66] This result demonstrates the tight connection between the actual hormone titer and feeding behavior of the larva. Therefore not only a phagodeterrent can affect the hormone titer via starvation[14] but *vice versa* the hormone titer can control feeding behavior.

The azadirachtin effect on molting processes and ecdysteroid titers has also been confirmed with other insect species. In *Rhodnius prolixus*, the substance inhibits ecdysis, if applied through a blood meal to fourth instar nymphs, with an effective dose (ED_{50}) of 4×10^{-4} µg/ml of blood. Here too, the authors explain feeding inhibition (ED_{50} = 25.0 µg/ml of blood) as an indirect effect due to an interference of azadirachtin with the endocrine system. Ecdysone given orally at 5.0 µg/ml and a juvenile hormone analog at 70 µg/insect counteracted the ecdysis inhibition as induced by azadirachtin.[57] Azadirachtins A and B induced equal effects. As in the *Epilachna* assay, the ED_{50} value for molt inhibition was higher for azadirachtin A (0.04) than for azadirachtin B (0.015 µg/ml). ATP, a phagostimulant, if added to the blood together with azadirachtin, reversed its antifeedant action.[42] The effects of azadirachtin A after injection into fourth instar nymphs of *Rhodnius prolixus* were studied on the level of molting inhibition, mortality, ecdysteroid titers, and mitotic index of cuticles. This study demonstrates the versatility of azadirachtin as a chemical probe for the insect endocrinologist. If injected 1 to 3 d after blood feeding, it inhibits the molting process. However, if injected during and after the onset of epidermal mitosis, ecdysis is not affected. A single dose blocks the onset of mitosis in the epidermis, which is associated with the molting cycle and is triggered by ecdysone. In the azadirachtin-treated group the ecdysteroid titers were too low for an induction of ecdysis. The authors prove that a single injection of as little as 2 ng of azadirachtin per nymph is enough for a 50% molt inhibition without any visible toxic effects during an extended larval period of more than 1 year! The half effective dose of mortality (40 ng/nymph) within 30 d is 20-fold higher than the dose needed for 50% molt inhibition. Even an ED_{90} with 10 ng/nymph is accompanied by not more than 5% mortality. It is notable, that after onset of the mitotic cycle azadirachtin no longer has any effect on ecdysis and as such does not block any essential receptors in the cuticle, which again contradicts the assumption of Käuser and Koolman[53] that azadirachtin is an antiecdysteroid.

A study by Dorn et al.[58] on the effects of azadirachtin on the molting cycle, endocrine system, and ovaries in last instar larvae of *Oncopeltus fasciatus* also supports the mutual and dose-dependent activity of this compound. As in *Locusta*, azadirachtin-induced permanent larvae exhibit adult characters. The epidermis of permanent larvae shows neither ecdysis nor apolysis; however, it engages in secretory activity which may correspond to adult procuticle secretion. These larvae show an ecdysteroid peak which is considerably delayed and distinctly lower than in the controls. With high azadirachtin doses adult ovarian development begins, in some cases leading to chorionated eggs. Also neurosecretion seems to be affected in these treated larvae and, possibly, the oviductal transformation from larval to adult form as well.[67] Likewise, in larvae of the tobacco hornworm, *Manduca sexta*, growth and endocrine events are affected by azadirachtin.[68] At low doses (0.5 µg/g), mainly defective pupae were formed, whereas with doses of 2 µg/g and higher mainly supernumerary larvae were formed. Most of these failed to cast off their exuvia and died after varying periods of survival. Here, also, the hemolymph ecdysteroid titers of azadirachtin-treated last instar larvae showed the characteristic shift is also reflected by a delayed larval growth. These larvae that succeeded in pupating, usually failed to perform normal pupal development. The hormonal effects are reflected by changes in the histological pattern of the prothroacic glands.[69] Similar results have been described for fifth instar larvae of *Bombyx mori*.[55] The authors also find a marked difference in azadirachtin effect regarding the age of insect at fifth stage. When 0- to 3-d-old larvae were injected, they were able to reach pupation but ultimately failed to undergo normal ecdysis and deformed pupae developed. If the larvae were treated on day 4 to 6, they showed characteristic black-band formation around their segments, followed by death in the larval stage, similar to the effects described for *Epilachna*.[3,70] This phenomenon was not seen, however, in the treated 0- to 3-d-old insects, albeit the rate of azadirachtin injection was the same for all larvae.

One can generalize from these findings that azadirachtin irreversibly, or at least for an extended period of time, blocks and sometimes changes developmental programs, as well as those which are normally expressed in the next instar only. Some examples are the adultoid characters of permanent larvae of *Locusta migratoria*,[62,63] of *Oncopeltus fasciatus*[58,67] and of *Manduca sexta*.[68] However, there are species-specific differences in the reaction from azadirachtin treatment, some insects not even reacting at all. Nothing is known about the molecular basis of these irreversible events. Tracer studies with tritiated 22,23-dihydroazadirachtin only showed that practically all of the excreted radioactivity was identical with the administered compound, and the same holds true for the radioactivity retained in the treated insect.[17]

B. EFFECT ON ENDOCRINE EVENTS OF THE ADULT INSECT

Inhibition of reproduction, either by crude neem kernel extracts or by azadirachtin, has long been known.[9,71-74] As in larvae, the main effect of azadirachtin is a change in the ecdysteroid titers.[48,56,67,68,75,76]

After a single injection of 10 μg azadirachtin into a female *Locusta migratoria* between days 2 and 13 after emergence, about 60% died during the following 4 d and all lost weight in the range of about 50%. If injected between days 2 and 10 (during the phase of previtellogenesis), no maturation of the terminal oocytes was observed. Injection between days 10 and 13 resulted in ovaries with almost mature oocytes. Similar to the effects in the larva, there is a sensitive phase during oocyte development. In the *Locusta* female this is obviously the period of vitellogenesis, whose program cannot be blocked by azadirachtin administration after its start. Most of the treated locusts had no oviposition and only traces of ecdysteroid were present in the ovaries.[71,72] Interestingly enough, similar effects can be induced by injection of antibrain antibodies. Such antibodies to brain material from mature *Locusta migratoria* females were raised in rabbits. *In vivo* injection of these antibodies into young locusts also inhibited ovarian development, indicating a possible blockade of allatotropic or gonadotropic activity.[77] The possibility of an involvement of allatotropic hormone[78] in the inhibitory activity of azadirachtin must be taken under consideration from these results and will be discussed later.

Similar azadirachtin effects are reported from other insects. If injected into newly hatched adults of *Oncopeltus fasciatus*, it affects longevity, fecundity, and hatchability of eggs from treated parents. There are marked differences between males and females, with higher mortality in the females and induction of impotence in the males with as little as 0.125 μg/insect, mainly due to mating failure which could be the consequence of an azadirachtin-induced hormonal imbalance in the males.[79] In *Schistocerca gregaria*, injection of 2 μg/g of the compound into newly hatched females completely inhibited weight gain within the first 8 d and then the females began growing again. Similar to body weight, no substantial increase in the wet weight of the ovary of the treated females was observed during the 12-d experimental period. Protein concentration in the hemolymph of the azadirachtin-treated females decreased continuously without any change in amino acid pool size during the first 8 d. In subsequent days, however, it was reduced considerably. Also, the hemolymph protein pattern changed in comparison with that of the controls. Azadirachtin delayed synthesis and release of neurosecretion from the A-type median neurosecretory cells of the brain, possibly leading to inhibition of ovarian development.[74]

What are the effects of azadirachtin on the ecdysteroid and juvenile hormone titers in the adult *Locusta migratoria* females, and how are these two hormones affected during the gonotrophic cycle? After a highly specific and sensitive technique for juvenile hormone quantification had become available,[80] this question could be approached in the living insect during its egg maturation cycle.[81] Every 24 h, 10 μl of hemolymph were collected through a puncture in the hindleg of the test locusts. Ecdysteroids, by radioimmunoassay, and juvenile hormone, by chemical derivatization and GC-MS-MIS measurement, were quantified in the same sample concomitantly. In some cases, vitellogenin was also quantified. It was possible, for the first time, with this technique to follow the time-dependent hormone shift after azadirachtin application and thereby to understand the hormonal regulation of oogenesis.

Juvenile hormone III is the only hormone analog present in *Locusta migratoria*.[82,83] This hormone controls synthesis of vitellogenin in the fat body and, as such, controls egg maturation in the ovary. A reduced juvenile hormone titer must therefore be followed by a reduced synthesis of egg yolk protein. The ecdysteroids are synthesized in the ovary at the end of the egg maturation cycle and, to some extent, then released into the hemolymph. Under normal control conditions, juvenile hormone and ecdysterone concentrations are low in the hemolymph during this so-called previtellogenesis after adult emergence. From day 5 onwards, the juvenile hormone titer increases steadily, reaching a maximum at the end of egg maturation and then decreasing steeply during oviposition. Due to the measurement of individuals it is possible to realize that, even in well-synchronized populations, there is a variation of the juvenile hormone peak maximum in the range of plus or minus 1 d, followed by a corresponding deviation in egg deposition. The juvenile hormone minimum during oviposition has also been described by Stay and Tobe.[84] Generally the juvenile hormone titer increases during previtellogenesis, reaching its maximum during rapid oocyte growth. The juvenile hormone peak is therefore a sensitive marker for the female reproductive period. Its duration may be only a very few hours[84] and the 24-h frame may therefore be too coarse. The same holds true for the ecdysteroid titers. Here again, the hemolymph hormone titer is tightly connected with the progress of egg maturation. A steep increase of ecdysteroid titer begins after the egg yolk protein has passed its maximum hemolymph titer and therefore an ecdysteroid peak becomes visible at the end of vitellogenesis. The two peaks of juvenile hormone and of ecdysone are more or less synchronous within the egg maturation cycles under normal conditions. Also the hemolymph peak of egg yolk protein coincides with the two hormone peaks. Application of azadirachtin could provide a better understanding of the correlation of these three components involved in egg maturation, i.e., juvenile hormone, ecdysteroids, and vitellogenin.

As previously discussed, azadirachtin interferes with developmental processes during insect metamorphosis.[48,55-58,68,71,72] Whereas these authors were interested primarily in the effect of ecdysteroid retardation, all the three parameters (juvenile hormone, ecdysterone, and vitellogenin) were now followed quantitatively. To make a long story very short: all these three factors are shifted concomitantly on the time axis after azadirachtin administration. What does that mean in terms of endocrine regulation? There is only one answer possible for the moment: all these three components are under the same neuroendocrine control of the *Pars intercerebralis*, at least as far as *Locusta migratoria* is concerned.

IV. CONCLUSIONS

With this argument in mind, not too many explanations of the mode of azadirachtin action are possible. All the facts speak in favor of an interference with the neuroendocrine control of metamorphosis. What could that mean? As already discussed, no structural homology with the ecdysteroids is possible due to the *cis, trans* isomerism of rings A and B in the two molecular structures. However, there are some similarities in the stereochemistry of all the azadirachtins. This has been discussed in connection with the minimum structural element for all the isomeric compounds. Due to the tracer experiments with tritiated dihydro-azadirachtin A, which is itself highly growth inhibitory, a group of hormonal events seems to be directly under neuroendocrine control. The amount of azadirachtin retained in the insect is extremely low and a trace is sufficient to trigger the many endocrine and, consequently, morphological and behavioral effects. The fact that in some cases, such as *Rhodnius*, such effects can last for the entire lifespan, indicates a high-affinity binding at a central part of the endocrine regulatory system. An extremely specific reaction of the azadirachtins inside the insect is therefore to be expected.[17] More and more indirect evidence for that assumption is coming up and has been discussed. There is the effect on juvenile and ecdysteroid titers, which can be abolished in the case of *Rhodnius*, whereas in other cases this is not possible. There is the expression of adult characters in permanent larvae. Or those effects which are expressed after only a successful molt. There is also the obvious involvement of other hormones like eclosion or diuretic hormone.[56]

As stated at the outset, we are now waiting for the fourth act, which must explain the molecular basis of azadirachtin action. What are, or is, the molecular target(s) in the insect? Not much is now known about high-affinity binding sites. A first result, indicating an interference of azadirachtin with protein biosynthesis, has been published recently by Fritzsche and Cleffmann.[85] In *Tetrahymena thermophila*, the compound, depending on the concentration applied, inhibits cell proliferation for a few hours. The cells then recover from this block to some extent. The initial inhibitory azadirachtin effect is strongest on RNA synthesis, which is reduced to 40% within the first 20 min. Then the rate of DNA synthesis decreases to about 30% of the control within 40 and 120 min. The authors conclude from their data that transcription products synthesized in the presence of azadirachtin are degraded at a high rate, which would explain the reduced cellular RNA contents. The effective dose of 10^{-6} *M* azadirachtin is in the same order of magnitude as the concentrations of other potent inhibitors effective in *Tetrahymena*. Some cellular events, therefore, should be very sensitive to azadirachtin.

REFERENCES

1. **Butterworth, J. H. and Morgan, E. D.,** Isolation of a substance that suppresses feeding in locusts, *Chem. Commun.,* 23, 1986.
2. **Zanno, P. R., Miura, I., Nakanishi, K., and Elder, D. L.,** Structure of the insect phago-repellent azadirachtin. Applications of PRFT/CWD carbon-13 nuclear magnetic resonance, *J. Am. Chem. Soc.,* 97, 1975, 1975.
3. **Schmutterer, H. and Rembold, H.,** The action of several pure fractions from seeds of *Azadirachta indica* on feeding and meta-morphosis of *Epilachna varivestis* (Col. Coccinellidae), *Z. Angew. Entomol.,* 89, 179, 1980.
4. **Butterworth, J. H. and Morgan, E. D.,** Investigations of the locust feeding inhibition of the seeds of the neem tree, *Azadirachta indica, J. Insect Physiol.,* 17, 169, 1971.
5. **Quadri, S. S. H. and Narsaiah, J.,** Effect of azadirachtin on the molting processes of last instar nymphs of *Periplaneta americana* (Linn.), *Indian J. Exp. Biol.,* 16, 1141, 1978.

6. **Ruscoe, C. N. E.,** Growth disruption effects of an insect antifeedant, *Nature (London), New Biol.,* 236, 159, 1972.

7. **Steets, R.,** The effect of crude extracts from the Meliaceae *Azadirachta indica* and *Melia azedarach* on various species of insects, *Z. Angew. Entomol.,* 77, 306, 1975.

8. **Steets, R.,** The Effect of Components from Meliaceae and Anacardiaceae on Coleoptera and Lepidoptera, Ph.D. thesis, University of Giessen, West Germany, 1976.

9. **Steets, R. and Schmutterer, H.,** Effect of azadirachtin on the life period and reproductive ability of *Epilachna varivestis* Muls. (Coleoptera, Coccinelidae), *Z. Pflanzenkr. Pflanzenschutz,* 82, 176, 1975.

10. **Meisner, J., Wysoki, M., and Ascher, K. R. S.,** The residual effect of some products from neem (*Azadirachta indica* A. Juss) seeds upon larvae of *Boarmia* (*Ascotis*) *selenaria* Schiff. in laboratory trials, *Phytoparasitica,* 4, 185, 1976.

11. **Ladd, T. L., Jr., Jacobson, M., and Buriff, C. R.,** Japanese beetle: extracts from neem tree seeds as feeding deterrents, *J. Econ. Entomol.,* 71, 810, 1978.

12. **Warthen, J. D., Jr.,** *Azadirachta indica:* a source of insect feeding inhibitors and growth regulators, U.S. Department of Agriculture, Science and Education Administration, ARM-NE-4, 1979.

13. **Rembold, H., Sharma, G. K., Czoppelt, Ch., and Schmutterer, H.,** Evidence of growth disruption in insects without feeding inhibition by neem seed fractions, *Z. Pflanzenkr. Pflanzenschutz,* 87, 290, 1980.

14. **Slama, K.,** The principles of antihormone action in insects, *Acta Entomol. Bohemoslov.,* 75, 65, 1978.

15. **Rembold, H. and Schmutterer, H.,** Disruption of insect growth by neem seed components, in *Regulation of Insect Development and Behavior,* Sehnal, F., Zabza, A., Menn, J. J., and Cymborowski, B., Eds., Wrozlaw Technical University Press, Wroczlaw, Poland, 1981, 1087.

16. **Forster, H.,** Isolation of Azadirachtins from Neem *(Azadirachta indica)* and Radioactive Labelling of Azadirachtin, M.S. thesis, University of Munich, West Germany, 1983.

17. **Rembold, H., Forster, H., Czoppelt, Ch., Rao, P. J., and Sieber, K. P.,** The azadirachtins, a group of insect growth regulators from the neem tree, in *Natural Pesticides from the Neem Tree and Other Tropical Plants,* Schmutterer, H. and Ascher, K. R. S., Eds., GTZ Press, Eschborn, West Germany, 1984, 153.

18. **Kundu, A. B., Ray, S., Chakrabarti, R., Nayak, L., and Chatterjee, A.,** Recent developments in the chemistry of C_{26}-terpen-

oids (tetranortriterpenoids), *J. Sci. Indian Res.,* 44, 256, 1985.

19. **Feuerhake, K. J. and Schmutterer, H.,** Simple procedures for isolating and formulating neem seed extracts and their action on various insects, *Z. Pflanzenkr. Pflanzenschutz,* 89, 737, 1982.

20. **Forster, H.,** Structure and Biological Action of Azadirachtins, a Group of Insect-Specific Growth Inhibitors from Neem (*Azadirachta indica*). Investigation of its Structure-Activity Effects on the Mexican Bean Beetle (*Epilachna varivestis*), Ph.D. thesis, University of Munich, West Germany, 1987.

21. **Turner, C. J., Tempesta, M. S., Taylor, R. B., Zagorski, M. G., Termini, J. S., Schroeder, D. R., and Nakanishi, K.,** An nmr spectroscopic study of azadirachtin and its trimethyl ether, *Tetrahedron,* 43, 2789, 1987.

22. **Uebel, E. C., Warthen, J. D., Jr., and Jacobson, M.,** Preparative reversed phase chromatographic isolation of azadirachtin from neem kernels, *J. Liq. Chromatogr.,* 2, 875, 1979.

23. **Warthen, J. D., Jr., Redfern, R. E., Uebel, E. C., and Mills, G. D., Jr.,** An antifeedant for fall armyworm larvae from neem seeds, U.S. Department of Agriculture, Science and Education Administration, ARR-NE-1, 1978.

24. **Warthen, J. D., Jr., Stokes, J. B., Jacobson, M., and Kozempel, M. F.,** Estimation of azadirachtin content in neem extracts and formulations, *J. Liq. Chromatogr.,* 7, 591, 1984.

25. **Yamasaki, R. B., Klocke, J. A., Lee, S. M., Stone, G. A., and Darlington, M. V.,** Isolation and purification of azadirachtin from neem (*Azadirachta indica*) seeds using flash chromatography and high-performance liquid chromatography, *J. Chromatogr.,* 356, 220, 1986.

26. **Schroeder, D. R. and Nakanishi, K.,** A simplified isolation procedure for azadirachtin, *J. Nat. Prod.,* 50, 241, 1987.

27. **Rembold, H., Forster, H., and Sonnenbichler, J.,** Structure of azadirachtin B, *Z. Naturforsch.,* 42C, 4, 1987.

28. **Bilton, J. N., Broughton, H. B., Jones, P. S., Ley, S. V., Lidert, Z., Morgan, E. D., Rzepa, H. S., Sheppard, R. N., Slawin, A. M. Z., and Williams, D. J.** An X-ray crystallographic, mass spectroscopic, and nmr study of the limonoid insect antifeedant azadirachtin and related derivatives, *Tetrahedron,* 43, 2805, 1987.

29. **Kraus, W., Bokel, M., Bruhn, A., Cramer, R., Klaiber, I., Klenk, A., Nagl, G., Pöhnl, H., Sadlo, H., and Vogler, B.,** Structure determination by nmr of azadirachtin and related compounds from *Azadirachta indica*

A. Juss (Meliaceae), *Tetrahedron*, 43, 2817, 1987.

30. **Klenk, A., Bokel, M., and Kraus, W.,** 3-Tigloylazadirachtol (tigloyl = 2-methylcrotonyl), an insect growth regulating constituent of *Azadirachta indica, J. Chem. Soc., Chem. Commun.,* 523, 1986.

31. **Kubo, I., Matsumoto, T., and Matsumoto, A.,** Structure of deacetylazadirachtinol. Application of 2D ^{1}H-^{1}H and ^{1}H-^{13}C shift correlation spectroscopy, *Tetrahedron Lett.,* 25, 4729, 1984.

32. **Rembold, H. and Forster, H.,** Structures of azadirachtins D—G, *Tetrahedron,* submitted.

33. **Kraus, W.,** Constituents of neem and related species. A revised structure of azadirachtin, in *New Trends in Natural Products Chemistry,* (Studies in Organic Chemistry, Vol. 26), Atta-ur-Rahman and Le Quesne, P. W., Eds., Elsevier, Amsterdam, 1986, 237.

34. **Butterworth, J. H., Morgan, E. D., and Percy, G. R.,** The structure of azadirachtin: the functional groups, *J. Chem. Soc. Perkin Trans.,* 2445, 1972.

35. **Rembold, H., Czoppelt, Ch., and Forster, H.,** Growth disrupting activity of the azadirachtin group, *Z. Angew. Entomol.,* in press.

36. **Noack, S. and Reichmuth, C.,** A mathematical procedure for determining the optional dose values of a preparation from empirically ascertained dose-effect data, *Mitt. Biol. Bundesans. Land. Forstwirtsch. Berlin-Dahlem,* 185, 1, 1978.

37. **Henderson, R., McCrindle, R., Melera, A., and Overton, K. H.,** Tetranortriterpenoids. IX. The constitution and stereochemistry of salannin, *Tetrahedron,* 24, 1525, 1968.

38. **Jacobson, M.,** Isolation and identification of insect antifeedents and growth inhibitors from plants: an overview, in *Natural Pesticides from the Neem Tree (Azadirachta indica A. Juss),* Schmutterer, H., Ascher, K. R. S., and Rembold, H., Eds., GTZ Press, Eschborn, West Germany, 1981, 13.

39. **Jacobson, M.,** The neem tree: natural resistance par excellence, *Natural Resistance of Plants to Pests,* American Chemical Society, Washington, D.C., 1986, 220.

40. **Jacobson, M., Stokes, J. B., Warthen, J. D., Jr., Redfern, R. E., Reed, D. K., Webb, R. E., and Telek, L.,** Neem research in the U.S. Department of Agriculture: an update, in *Natural Pesticides from the Neem Tree and Other Tropical Plants,* Schmutterer, H. and Ascher, K. R. S., Eds., GTZ Press, Eschborn, West Germany, 1984, 31.

41. **Schmutterer, H.,** Neem research in the Federal Republic of Germany since the First International Neem Conference, in *Natural Pesticides from the Neem Tree and Other Tropical Plants,* Schmutterer, H. and Ascher, K. R. S., Eds., GTZ Press, Eschborn, West Germany, 1984, 21.

42. **Garcia, E. S., Azambuja, P., Forster, H., and Rembold, H.,** Feeding and moult inhibition by azadirachtins A, B, and 7-acetylazadirachtin A in *Rhodnius prolixus* nymphs, *Z. Naturforsch.,* 39C, 1155, 1984.

43. **Meisner, J., Ascher, K. R. S., Aly, R., and Warthen, J. D., Jr.,** Response of *Spodoptera littoralis* (Boisd.) and *Earias insulana* (Boisd.) larvae to azadirachtin and salannin, *Phytoparasitica,* 9, 27, 1981.

44. **Kubo, I. and Klocke, A.,** Azadirachtin, insect ecdysis inhibitor, *Agric. Biol. Chem.,* 46, 1951, 1982.

45. **Sharma, G. K., Czoppelt, Ch., and Rembold, H.,** Further evidence of insect growth disruption by neem seed fractions, *Z. Angew. Entomol.,* 90, 439, 1980.

46. **Shinfoon, C., Xing, Z., Siuking, L., and Duanping, H.,** Growth-disrupting effects of azadirachtin on the larvae of the Asiatic corn borer (*Ostrinia furnacalis* Guenee), *Z. Angew. Entomol.,* 95, 276, 1985.

47. **Gill, J. S.,** Studies on Insect Feeding Deterrents with Special Reference to the Fruit Extracts of the Neem Tree, *Azadirachta indica,* A. Juss., Ph.D. thesis, University of London, England, 1972.

48. **Redfern, R. E., Warthen, J. D., Jr., Uebel, E. C., and Mills, G. D., Jr.,** The antifeedant and growth-disrupting effects of azadirachtin on *Spodoptera frugiperda* and *Oncopeltus fasciatus,* in *Natural Pesticides from the Neem Tree (Azadirachta indica A. Juss),* Schmutterer, H., Ascher, K. R. S., and Rembold, H., Eds., GTZ Press, Eschborn, West Germany, 1981, 129.

49. **Rembold, H. and Czoppelt, Ch.,** Assay of plant-derived inhibitors of insect growth from *Azadirachta indica* in the assay test of bee larvae, *Mitt. Dtsch. Ges. Allgem. Angew. Entomol.,* 3, 196, 1981.

50. **Rembold, H., Sharma, G. K., and Czoppelt, Ch.,** Growth regulating activity of azadirachtin in two holometabolous insects, in *Natural Pesticides from the Neem Tree (Azadirachta indica A. Juss),* Schmutterer, H., Ascher, K. R. S., and Rembold, H., Eds., GTZ Press, Eschborn, West Germany, 1981, 121.

51. **Gaaboub, I. and Hayes, D. K.,** Biological activity of azadirachtin, component of the neem tree inhibiting molting in the face fly, *Musca autumnalis* de Geer, *Environ. Entomol.,* 13, 803, 1984.

52. **Gaaboub, I. and Hayes, D. K.,** Effect of larval treatment with azadirachtin, a molting inhibitory component of the neem tree, on reproductive capacity of the face fly, *Musca autumnalis* de Geer, *Environ. Entomol.,* 13,

1639, 1984.

53. **Käuser, G. and Koolman, J.,** Ecdysteroid receptors in tissues of the blowfly, *Calliphora vicina,* in *Advances in Invertebrate Reproduction,* Vol. 3, Engels, W., Ed., Elsevier, Amsterdam, 1984, 602.

54. **Spindler, D.,** personal communication.

55. **Koul, O., Amanai, K., and Ohtaki, T.,** Effect of azadirachtin on the endocrine events of *Bombyx mori, J. Insect Physiol.,* 33, 103, 1987.

56. **Sieber, K. P. and Rembold, H.,** The effects of azadirachtin on the endocrine control of moulting in *Locusta migratoria, J. Insect Physiol.,* 29, 523, 1983.

57. **Garcia, E. S. and Rembold, H.,** Effects of azadirachtin on ecdysis of *Rhodnius prolixus, J. Insect Physiol.,* 30, 939, 1984.

58. **Dorn, A., Rademacher, J. M., and Sehn, E.,** Effects of azadirachtin on the moulting cycle, endocrine system, and ovaries in last-instar larvae of the milkweed bug, *Oncopeltus fasciatus, J. Insect Physiol.,* 32, 231, 1986.

59. **Zebitz, C. P. W.,** Effect of some crude and azadirachtin-enriched neem (*Azadirachta indica*) seed kernel extracts on larvae of *Aedes aegypti, Entomol. Exp. Appl.,* 35, 11, 1984.

60. **Sieber, K. P.,** Investigations of the Effect of Azadirachtin on the Development and its Regulation in *Locusta migratoria,* Ph.D. thesis, University of Munich, West Germany, 1982.

61. **Rembold, H.,** Secondary plant products in insect control, with special reference to the azadirachtins, in *Advances in Invertebrate Reproduction,* Vol. 3, Engels, W., Ed., Elsevier, Amsterdam, 1984, 481.

62. **Shalom, U. and Pener, M. P.,** Sexual behavior without adult morphogenesis in *Locusta migratoria, Experientia,* 40, 1418, 1984.

63. **Kutsch, W.,** Pre-imaginal flight motor pattern in *Locusta, J. Insect Physiol.,* 31, 581, 1985.

64. **Han, S. Z.,** Effect of Azadirachtin on the Circadian Rhythm of the Cockroach *Leucophaea maderae* (Fabricius) Locomotion, Ph.D. thesis, University of Tübingen, West Germany, 1986.

65. **Han, S. Z. and Engelmann, W.,** Azadirachtin affects the circadian rhythm of locomotion in *Leucophaea maderae,* submitted.

66. **Rao, P. J. and Rembold, H.,** Effect of ecdysterone on food intake of *Locusta migratoria* hoppers, *Z. Naturforsch.,* 38C, 878, 1983.

67. **Dorn, A., Rademacher, J. M., and Sehn, E.,** Ecdysteroid-dependent development of the oviduct in last-instar larvae of *Oncopeltus fasciatus, J. Insect Physiol.,* 32, 643, 1986.

68. **Schlüter, U., Bidmon, H. J., and Grewe, S.,** Azadirachtin affects growth and endocrine events in larvae of the tobacco horn-worm, *Manduca sexta, J. Insect Physiol.,* 31, 773, 1985.

69. **Bidmon, H. J.,** Ultrastructural changes in the prothoracic glands of *Manduca sexta* larvae (Lepidoptera) untreated and treated with azadirachtin, *Entomol. Gener.,* 12, 1, 1986.

70. **Schlüter, U.,** Histological observations on the phenomenon of black legs and thoracic spots: effects of pure fractions of neem kernel extracts on *Epilachna varivestis,* in *Natural Pesticides from the Neem Tree (Azadirachta indica A. Juss),* Schmutterer, H., Ascher, K. R. S., and Rembold, H., Eds., GTZ Press, Eschborn, West Germany, 1981, 97.

71. **Rembold, H. and Sieber, K. P.,** Inhibition of oogenesis and ovarian ecdysteroid synthesis by azadirachtin in *Locusta migratoria, Z. Naturforsch.,* 36C, 466, 1981.

72. **Rembold, H. and Sieber, K. P.,** Effect of azadirachtin on oocyte development in *Locusta migratoria migratorioides,* in *Natural Pesticides from the Neem Tree (Azadirachta indica A. Juss),* Schmutterer, H., Ascher, K. R. S., and Rembold, H., Eds., GTZ Press, Eschborn, West Germany, 1981, 75.

73. **Koul, O.,** Azadirachtin. II. Interaction with the reproductive behaviour of red cotton bugs, *Z. Angew. Entomol.,* 98, 221, 1984.

74. **Subrahmanyam, B. and Rao, P. J.,** Azadirachtin effects on *Schistocerca gregaria* Forskal during ovarian development, *Curr. Sci.,* 55, 534, 1986.

75. **Garcia, E. S., Uhl, M., and Rembold, H.,** Azadirachtin, a chemical probe for the study of moulting processes in *Rhodnius prolixus, Z. Naturforsch.,* 41C, 771, 1986.

76. **Mordue (Luntz), A. J., Evans, K. A., and Charlett, M.,** Azadirachtin, ecdysteroids and ecdysis in *Locusta migratoria, Comp. Biochem. Physiol.,* 85C, 297, 1986.

77. **Rembold, H., Eder, J., and Ulrich, G. M.,** Inhibition of allatotropic activity and ovary development in *Locusta migratoria* by anti-brain-antibodies, *Z. Naturforsch.,* 35C, 1117, 1980.

78. **Ulrich, G. M., Schlagintweit, B., Eder, J., and Rembold, H.,** Elimination of the allatotropic activity in locusts by microsurgical and immunological methods: evidence for humoral control of the corpora allata, hemolymph proteins, and ovary development, *Gen. Comp. Endocrinol.,* 59, 120, 1985.

79. **Dorn, A.,** Effects of azadirachtin on reproduction and egg development of the heteropteran *Oncopeltus fasciatus* Dallas, *J. Appl. Entomol.,* 102, 313, 1986.

80. **Rembold, H. and Lackner, B.,** Convenient method for the determination of picomole amounts of juvenile hormone, *J. Chromatogr.,* 323, 355, 1985.

81. **Uhl, M.,** Experimental Alteration of the Ecdysteroid and Juvenile Hormone Titer During

the Egg Ripening Cycle of *Locusta migratoria*, Ph.D. thesis, University of Munich, West Germany, 1987.

82. **Bergot, B. J., Schooley, D. A., and de Kort, C. A. D.,** Identification of JH III as the principal juvenile hormone in *Locusta migratoria, Experientia,* 37, 909, 1981.

83. **Rembold, H.,** Modulation of JH III titer during the gonotrophic cycle of *Locusta migratoria,* measured by gas chromatography-selected ion monitoring mass spectrometry, in *Juvenile Hormone Biochemistry,* Pratt, G. E. and Brooks, G. T., Eds., Elsevier/North Holland, Amsterdam, 1981, 11.

84. **Stay, B. and Tobe, S. S.** Structure and regulation of the corpus allatum, *Adv. Insect Physiol.,* 18, 304, 1985.

85. **Fritzsche, U. and Cleffmann, G.,** The insecticide azadirachtin reduces predominantly cellular RNA in *Tetrahymena, Naturwissenschaften,* 74, 191, 1987.

4

Effects of Neem on Pests of Vegetables and Fruit Trees

Heinrich Schmutterer and Carsten Hellpap
Institute of Phytopathology and Applied Zoology
Justus-Liebig-University
Geissen, West Germany

TABLE OF CONTENTS

I. INTRODUCTION

Some active ingredients of the seeds and leaves of the tropical neem tree, *Azadirachta indica,* especially the tetranortriterpenoid azadirachtin, influence the feeding behavior, metamorphosis (IGR-effect), fecundity, and fitness of numerous insect species belonging to various orders. Some spider mites are also affected. Apart from some smaller, less important groups other insect orders, namely Saltatoria, Homoptera, Heteroptera, Coleoptera, Lepidoptera, Hymenoptera, and Diptera proved to be sensitive, at least in laboratory experiments. These orders comprise numerous important vegetable and fruit-tree pests. Various nematode species are also influenced by neem products, provided they are worked into the soil.

During recent years a number of neem products have been applied in laboratory and field trials to test their suitability for practical pest control purposes. Preference was given by most research workers to studies of neem effects on vegetable insects, whereas fruit-tree pests were somewhat neglected.

In this article, results of field experiments will be listed and discussed first as they are more important than laboratory trials in the light of practical pest control by application of neem products. It is a well-known fact that good or even excellent results in the laboratory do not mean automatically that the same has to be expected under field conditions, especially in case of insect growth regulating (IGR) substances, such as azadirachtin.

II. NEEM PRODUCTS

It is essential to give a brief review of the most important neem products which have been and are being used for insect, mite, and nematode control experiments.

The pure, active substance, azadirachtin, is exclusively used in the laboratory, usually for physiological experiments. It is too expensive to extract and purify sufficient amounts of this compound for pest control purposes and its synthesis is also not economically feasible for the time being.

The most important neem products used in numerous trials are as follows: aqueous neem seed kernel and leaf extracts ("suspensions"); alcoholic (ethanolic, methanolic) seed kernel and leaf extracts; enriched, formulated seed kernel extracts; neem seed oil and neem seed cake.

Aqueous extracts ("suspensions") are prepared from ground or pounded dried neem seeds or dried leaves. The former are often preferred as they are more active. Usually about 25 to 50 g of seed kernels are used per liter of water. If unshelled seeds are used, double is needed. The extraction process lasts 5 to 6 h. After that, the larger particles are removed by filtering the liquid through a piece of cloth and the spray is ready for application at a high volume.[1]

One-step alcoholic (ethanolic, methanolic) seed kernel or leaf extracts have also been used but their production is more expensive and also more time-consuming than that of water extracts. Alcoholic extracts can be obtained by extraction of ground or pounded seed kernels or leaves in the cold or by means of Soxhlet apparatus. For successful pest control, concentrations of a few percent (1 to 2% for seed extracts, ~2 to 4% for leaf extracts) are needed.[2-4]

Enriched, formulated seed kernel products are either based on alcoholic extracts, which are purified in some steps to increase their azadirachtin content, as in the case of the U.S.-patented "Margosan-O",[5] or extracts obtained with an azeotropic mixture of methanol and methyl tertiary-butyl ether, also purified in some steps, as in the case of

AZT-VR-K.[6,7] There are some other enriched products which have been developed in India and elsewhere during recent years, but field experiments in vegetables and fruit trees with these pesticides are still rather scarce or published in journals which are not easily accessible. Formulations of enriched products are made by addition of various emulsifiers or sometimes of neem oil.

Neem seed oil may be pressed from neem seeds in the cold by using oil presses or by extraction with alcohols or other solvents using Soxhlet apparatus. Small amounts of oil can be obtained by kneading neem seed powder by hand after adding some water.[1] Neem kernels contain up to 50% of oil.

Neem oil is emulsified by addition of emulsifiers. It may be sprayed, after mixing with water, at concentrations of a few percent (e.g., 3 to 5%) or at higher concentrations which may lead to phytotoxicity in some plant species.

Neem seed cake is the remains of seed kernels after pressing the oil therefrom. This cake is a useful organic fertilizer, containing several percent of nitrogen as well as some azadirachtin and/or other active principles. It also has nitrification properties. This cake, worked into the soil in various concentrations, proved to be effective against some soil- and root-infesting nematodes. Nimbin, deacetylnimbin, and thionemon were, *inter alia,* identified as active principles against this group of pests.[8]

Fresh and dried neem leaves are also somewhat effective against nematodes after they have been worked into the soil, but considerable amounts are needed to obtain satisfactory results.

III. EXPERIMENTAL RESULTS OF INSECT PESTS AND MITES ON VEGETABLES AND FRUIT TREES

A. VEGETABLES

1. Cabbage, Cauliflower, Chinese Kale, and Other Crucifers

In Togo, cabbage field plots heavily infested with the diamondback moth, *Plutella xylostella,* were treated with different neem products at intervals of 7 d starting 5 weeks after transplanting.[2] A water extract of powdered neem leaves (suspension) (40 g/l) gave no protection in comparison to the control. On the other hand, a neem leaf methanolic extract at 4% significantly increased the yield and reduced the number of damaged heads and of larvae per infested head. The best results were obtained with a methanolic extract of neem kernels. At both concentrations tested (2 and 4%) yields were considerably higher than in the control. There was no significant difference between the plots sprayed with the synthetic insecticides mevinphos (0.05%) or deltamethrin (0.02%) and those sprayed with the neem kernel extracts.

Aqueous extracts of unshelled neem seeds showed a high efficacy against *P. xylostella* larvae in other field tests in Togo. Sprays based on 25 and 50 g of neem powder per liter of water prevented any feeding damage by the cabbage moth and also controlled the oriental cabbage webworm, *Hellula undalis,* very well. Consequently, the number of cabbage heads harvested, the quality, and the weight per head were considerably higher in the neem plots. Dipel, a microbiological product based on *Bacillus thuringiensis,* proved to be considerably less active than the neem extracts.[1]

In a field test in Thailand, an aqueous neem seed kernel extract at 3.33% (100 g/3 l water) and an enriched water extract called "Nm1" at 3.33 and 1.67% were tested against *P. xylostella* on Chinese kale. Nm1 was obtained by extracting 100 g of neem seed kernels in methanol and then extracting the residue in 3 l of water. The aqueous extract was mixed with petroleum ether and separated. The final concentration of the prepared extract was defined as 3.33%. The Chinese kale plants were sprayed five

Table 1
EFFECT OF NEEM EXTRACT AND CONVENTIONAL INSECTICIDE TREATMENTS ON CABBAGE YIELD AND HEAD DAMAGE[10]

Insecticide, conc or applied amount	Yield (t/ha)	Head damage*
Control	34.6 a**	4.9 a
Selecron 500 EC, 0.05% a.i.	42.3 ab	4.9 a
BAY SIR 14591 250 EC, 50 g a.i./ha	71.1 c	1.0 d
Thuricide HP, 1 kg prod./ha	34.0 a	4.8 a
AZT-VR-K EC, 0.20%	52.9 b	2.6 b
AZT-VR-K EC, 0.20% + Thuricide HP, 1 kg prod./ha	47.3 ab	2.3 b
ANSKE, 25 g NSK/1	45.5 ab	1.5 c
ANSKE, 25 g NSK/1 + Thuricide HP, 1 kg prod./ha	44.3 ab	1.4 b

*   Damage rating: 1 = heads with no damage; 2 = heads with light damage, no trimming required; 3 = heads with light damage, trimming of one leaf required; 4 = heads with moderate damage, trimming of two to three leaves required; 5 = heads with severe damage, trimming of more than three leaves required.

** Within columns, followed by a common letter are not significantly different at p = 0.05 (Duncan's Multiple Range Test).

times, once every 5 d. All neem treatments reduced considerably the number of larvae of *P. xylostella* and *Spodoptera litura.* The application of cypermethrin at 0.025% was slightly less effective than that of Nm1 (3.33%) but the difference was not significant. The addition of piperonyl butoxide considerably increased the efficacy of the neem extracts.[9]

In the Philippines, several field tests were carried out to test the efficacy of various neem products against *P. xylostella.* In these trials aqueous and enriched alcoholic seed kernel extracts (AZT-VR-K) alone or in combination with the *Bacillus thuringiensis*-based product, Thuricide, were compared to each other and to the synthetic insecticides Selecron (profenofos) and BAY SIR 14591 (IGR, developed by Bayer AG). The cabbage plants were treated 5 times starting 11 d after transplanting. The order of effectiveness with reference to the yield was BAY SIR 14591 > AZT-VR-K (0.2%) > AZT-VR-K (0.2%) + Thuricide > aqueous kernel extract (25 g/l) > aqueous kernel extract (25 g/l) + Thuricide > Selecron > Thuricide. The development of the larvae in the treated neem plots was seriously affected so that considerably lower numbers of last instar larvae, prepupae, and pupae could be found. Due to heavy infestation pressure the feeding activity of larvae could not be prevented completely in the neem plots, but most of the cabbage heads showed only light, tolerable damage (Table 1).[10]

Laboratory trials in West Germany confirmed that the efficacy of neem extracts in the control of insect pests of cabbage and other crops can be considerably increased by adding piperonyl butoxide (p.b.) to the extracts. In general, ratios of 10:1 to 5:1 (p.b.:extract) were necessary to get high synergistic or additive effects. Tested larvae of *P. xylostella* and the Colorado potato beetle, *Leptinotarsa decemlineata*, showed higher mortalities after uptake of the neem-p.b. mixture in comparison to the neem treatments alone. In some cases also an acceleration of the time necessary to kill a certain percentage of insects was observed.[11] In further studies it could be demonstrated that other synergists like Tropital and Sesoxane are even more effective than p.b. and that a ratio of 2:1 (p.b.:neem extract) is not sufficient to produce an increase of pest mortalities.[12]

In field tests in Mauritius against the cabbage webworm, *Crocidolomia binotalis*, finely ground or oven-dried leaves of neem were stirred for about 24 h in 95% ethanol. From the dried crude neem leaf extract a 2% solution was prepared which was applied weekly to cabbage, starting from the third week after transplantation of seedlings. The neem treatment reduced the number of infested plants by about 40% in comparison with the control. Egg-laying by adult moths was reduced by about 60%. Neem proved to be as effective as the insecticide Ultracide and *Bacillus thuringiensis*.[13] Prior laboratory tests showed that methanolic extracts of neem leaves gave higher yields and were more effective than ethanolic and aqueous extracts and therefore may also give slightly better results in the field.[14]

In further studies in Mauritius, the efficacy of neem extracts to control cabbage pests was compared to that of Decis (deltamethrin). Cabbage plants were sprayed in the field with 2.5% of an acetonic neem seed kernel extract once a week starting 2 weeks after transplanting of seedlings. The neem treatment reduced the number of larvae of the two major lepidopterous pests, *P. xylostella* and *C. binotalis*, and consequently the percentage of infested plants per plot (neem about 12%, control about 65%). The performance of Decis was slightly but not significantly better than that of neem.[15]

The same experimental design in Chinese cabbage and cauliflower confirmed that *P. xylostella* can efficiently be controlled by neem. In the neem plots the incidence of Chinese cabbage or cauliflower damage was about one third compared to that in the control due to a reduction of the population of *P. xylostella*. In both crops the results of neem and Decis were not significantly different.[15]

Neem substances in the U.S. gave good protection against another serious cabbage pest, the cabbage looper *Trichoplusia ni.* Sprayed weekly on cabbage in the field, 2% of an ethanolic seed extract had an efficacy of about 80% against larvae whereas Ammo, a standard insecticide, killed more than 95% of the pest population.[16]

In the People's Republic of China, 2% neem oil completely inhibited feeding of fifth instar larvae of the imported cabbageworm, *Pieris rapae,* in laboratory tests.[17]

In the Philippines, the application of various neem seed kernel extracts on cabbage in the field successfully controlled larvae of *P. canidia.* Treatments with the enriched extracts AZT-VR-K EC at 0.1% and MTB/H_2O-K at 0.1% showed an efficacy of over 85% referring to the number of larvae. The additionally tested insect growth regulating insecticide BAY SIR 14591 gave similar results. The application of a methanolic extract of neem kernels at 0.1% halved the number of larvae and pupae of *P. canidia.*[18]

In field studies on cabbage in West Germany, neem seed extracts based on various organic solvents and neem oil diminished considerably the number of larvae of *P. brassicae.* It could be demonstrated that different formulations have a strong influence on the efficacy of the neem extracts and oil and on the period during which the neem products are effective (residual effect).[19]

In a field test in India the efficacy of different neem products was evaluated against insect pests of radish. An aqueous seed kernel extract, ethanolic seed kernel extract, and neem oil, all at concentrations of 1, 2, and 3%, respectively, were sprayed on young radish plants 7, 17, 27, and 37 d after sowing. All neem treatments reduced leaf damage by the flea beetle, *Phyllotreta downsei.* The best results were obtained with aqueous seed kernel extract, giving between 60 and 68% protection, followed by ethanolic extract and neem oil.[20]

Aqueous neem seed kernel extracts and an AZT-VR-K extract sprayed on young cabbage plants could not significantly reduce feeding damage by a flea beetle species of the genus *Phyllotreta* in the Philippines.[18]

2. Cucurbits

In field trials in Gambia, a weekly application of an aqueous extract of ripe neem fruits ("berries") at 112.5 g/l and a fortnightly application of a neem fruit water suspension (85 g/l) controlled efficiently the African melon ladybird, *Henosepilachna elaterii* (=

Epilachna chrysomelina). The number of cucumbers per plant was higher than in the malathion treatment (85 ml/4.5 1).[21]

In field tests in Togo, aqueous extracts of neem seed kernels (50 and 10 g/l) and neem oil (10 and 20 l/ha) did not affect the feeding behavior and oviposition of adults of *H. elaterii* on zucchini. However, all neem treatments reduced considerably or prevented feeding by larvae of this insect.[22]

The effects of azadirachtin and salannin on the striped cucumber beetle, *Acalymma vittatum*, and the spotted cucumber beetle, *Diabrotica undecimpunctata*, were tested in laboratory and greenhouse trials in the U.S. Leaf disks, dipped in an azadirachtin solution of 0.01%, stopped feeding by *A. vittatum* whereas 0.1% of azadirachtin was necessary to reduce foliar damage by *D. undecimpunctata* to 98%. Salannin was also very active against the former but less effective against the latter. Further greenhouse trials showed that azadirachtin concentrations of 0.1% gave good protection against *A. vittatum* for 3 d. To extend the effect for a longer period and to reduce transmission of diseases by the beetles, concentrations of at least 0.5% had to be applied. In greenhouse tests it was demonstrated that azadirachtin can act systemically when the roots of plants are immersed in azadirachtin solutions.[23]

In Togo, water extracts of unshelled neem seeds (50 g/l water and 100 g/l) and neem oil (10 and 20 1/ha) were sprayed weekly for 3 weeks to zucchini plots in the field, starting 21 d after transplantation and then twice a week during the main fruiting phase of the plants. The water extract was applied with a knapsack sprayer, the neem oil with a ULV sprayer. In all neem treatments the number of juvenile instars per leaf of the tobacco whitefly, *Bemisia tabaci*, was reduced by about 70%. There was no significant difference between the various neem treatments.[22] Some further results against *B. tabaci* were obtained in laboratory tests in the U.S.; 0.2 and 2% of an ethanolic extract of neem seeds increased egg and larval mortalities and reduced oviposition.[24]

Aqueous extracts of unshelled neem seeds (50 and 100 g/l) and neem oil (10 and 20 1/ha, ULV) were sprayed in Togo against the fruit fly, *Dacus ciliatus*, on zucchini in the field. The neem treatments could not reduce the attack by the pest and the degree of damage to the plants.[22] Antiovipositional effects against *Dacus* spp. on vegetables were observed in India; 2.5% of an ethanolic extract of neem oil, extracted with petroleum ether, deterred egg-laying of *D. cucurbitae*. However, a concentration of 20% was necessary to obtain the same effect with *D. dorsalis*.[25]

3. Beans (Common Bean, Broad Bean)

Laboratory tests in Israel showed that neem seed kernel extracts with nonpolar solvents are more effective than with polar ones in the control of the carmine spider mite, *Tetranychus cinnabarinus*, a pest of beans and other vegetables. The best results were obtained with pentane, followed by chloroform, *n*-butanol, acetone, methanol, and water. Adult females were repelled from treated bean leaves and laid fewer eggs. However, the chloroform and butanol extracts caused phytotoxic effects. Adult females of the mite, when sprayed directly with a pentane or an acetone extract, showed reduced fecundity and high mortality.[26]

In Tanzania, the effectiveness of neem extracts to control several bean pests was compared to that of other plant extracts (tomato, hot pepper) and the synthetic insecticide Lindane. Aqueous extracts of neem seed kernels at 2% and of neem leaves at 4%, sprayed in the field on bean plants, halved the incidence and the damage of the flower thrips, *Taeniothrips sjostedti*. In the neem treatments the number of larvae of the pod borers, *Maruca testulalis* and *Heliothis armigera,* and the pod sucking bug, *Acanthomia horrida*, were also reduced (up to 40%). The seed extract proved to be slightly more active than the leaf extract. The results of the Lindane plots were not significantly different from those of the neem plots.[27]

Aphids are, in general, not very susceptible to foliar applications of neem. However, laboratory trials in West Germany with *Vicia faba* and the aphid species *Acyrthosiphon pisum* and *Aphis fabae* demonstrated that high concentrations of methyl tertiary-butyl ether (MTB) extracts and methanol = methyl tertiary-butyl ether (AZT) extracts of neem seed kernels can cause high mortalities and strongly affect development and fecundity of these homopterans. The MTB extract considerably increased the mortality rates. A further improvement of the efficacy was achieved by adding lecithins and sesame oil to this mixture.[28]

Several laboratory studies in West Germany revealed that the Mexican bean beetle, *Epilachna varivestis* is highly susceptible to methanolic neem seed extracts and azadirachtin,[3,29-31] However, there are presently no field trials reported. It may be concluded from laboratory trials that an aqueous extract ("suspension") of neem seed kernels (25 to 50 g/l) or a methanolic extract at 2% will give satisfactory control of this bean and soybean pest under field conditions in Central and North America.

In a series of field trials in Tanzania the application of various neem seed kernel and leaf extracts in comparison with the synthetic insecticide, dimethoate, gave promising results in controlling the adults of the foliar beetle, *Ootheca bennigseni*, an important pest of beans and other legumes. However, none of the treatments, including the standard insecticide, protected the bean plants completely. Referring to the degree of foliar damage by the beetle, the applications of a 4% aqueous neem seed kernel extract and dimethoate were most effective, followed by 2% aqueous and 2% alcoholic neem seed kernel extracts. The aqueous leaf extract at 2% and seed kernel dust were less active and only slightly better than the control.[32]

4. Cowpea and Other Legumes

In a field test in Togo 4 and 2% methanolic neem leaf and seed powder extracts, respectively, were tested against several pests (Lepidoptera, Coleoptera, Homoptera) of cowpea, *Vigna unguiculata*. In plots treated with neem seed extracts most pest species could be controlled so that the neem treatments gave higher yields than the control plots. However, the yields were lower than in the plots sprayed with the broad-spectrum standard insecticide, endosulfan, at 1.5%.[33]

A good control effect of neem against *H. armigera* on legumes was also confirmed by several other field trials in India. For instance, neem seed extract (50 g/l of water) was very effective in reducing the population of *H. armigera* larvae on gram, *Cicer arietinum*, and increased the yield by about 40%. Neem was comparable in its activity to eight tested conventional insecticides.[34]

Applications of aqueous neem seed kernel extract ("suspension") in India at 1, 5, and 8% on pigeon pea, *Cajanus cajan*, controlled efficiently the pod fly, *Melanagromyza obtusa*, and the pod borer, *Heliothis armigera*.[35,36] The efficacy was comparable to that of synthetic pesticides. Similar results were obtained with neem oil and ethanolic extracts of seeds against *M. obtusa*, *H. armigera*, and *Maruca testulalis* on the same crop.[37]

In field trials in India, 1% concentrations of a water-dispersible neem seed powder and a dust produced from crushed kernels reduced the damage caused by the butterfly, *Euchrysops cnejus*, significantly and increased the yield of *Vigna radiata* (green gram). The performance of these products was comparable to that of synthetic pyrethroids, viz., flucythrinate, fenvalerate, and cypermethrin, as well as to other insecticides, such as quinalphos, malathion, and endosulfan.[38]

The hairy caterpillar, *Amsacta moorei*, is a serious polyphagous pest of several crops in India. In field tests the application of aqueous neem seed kernel extracts (1 g/l water and 10 g/l) on sunhemp, *Crotolaria juncea*, caused complete mortality after 2 d.[39] These results were confirmed by semifield studies in which dew-gram plants, *V. aconitifolia*, were treated with aqueous "suspensions" of neem seed kernels in the field and than offered in the laboratory to fifth instar larvae of *A. moorei*. A concentration of 0.1% (1 g/l water) prevented nearly any damage by the hairy caterpillars shortly after application. However, long residual effects (up to 7 d) could only be achieved at concentrations of 1%. An aqueous neem kernel extract ("suspension") of 0.5% (5 g/l) showed systemic effects against *A. moorei* when applied to the soil of potted dew-gram plants.[40]

Neem products were successfully used against the vegetable leaf miner, *Liriomyza sativae*, in laboratory tests in the U.S. Margosan-O, a commercial product based on a neem kernel extract and azadirachtin caused high mortalities of larvae and pupae of this leaf miner species when applied to the soil or aqueous nutrient solutions of several crops. On tomato plants the efficacy of neem was lower than on lima bean plants.[41,42] When sprayed on the latter, a 0.1% ethanolic extract of neem seed kernels killed almost all of the tested first, second, and third instar larvae of *L. sativae*. The egg-stage was not affected by neem. However, the neem treatment was effective against freshly emerging larvae up to 7 d after application. In this period the effectiveness declined from 86% (0 d) to 44% (7th d).[43]

Ethanolic extracts of neem seeds applied to leaves of lima beans caused 100% mortality of the larvae of the onion leaf miner, *L. trifolii*, shortly after emergence.[43]

Apart from extracts of the seeds, neem leaf extracts can also be used to control *L. trifolii*. In Mauritius, a leaf extract, obtained by steam distillation and separation of the oily parts with acetone, was tested in the laboratory. A 2% aqueous solution of that extract reduced the number of live pupae and flies per infested bean leaf by about 75 and 80%, respectively. However, the neem treatment also killed about 90% of the natural hymenopteran parasite, *Hemiptarsenus semialbiclave,* which was abundant in the test. Furthermore, the neem leaf extract proved to be phytotoxic to young cotyledonous bean leaves.[44]

5. Okra

Methanolic neem seed kernel extracts at concentrations of 1, 2, and 4% and a methanolic neem leaf extract at 4% were applied in Togo to okra plants in the field to control adults of the flea beetles, *Podagrica* spp., and the larvae of the cotton leaf-roller, *Sylepta derogata*. The plots were sprayed six times at an interval between the treatments of 7 d, starting 3 weeks after germination. All neem treatments reduced considerably the number of leaves with feeding holes of *Podagrica* spp. per plant. However, the neem extracts could not keep the plants free from the beetles so that some damage occurred. The application of deltamethrin at 0.2% reduced the number of beetles by more than 90% and consequently was more efficient in the control of these pests.[45]

Methanolic neem seed extracts at 4 and 1% proved to be highly effective against *S. derogata*. In these neem treatments the percentage of leaves rolled by the larvae was more than 90 times lower than in the control plots. The efficacy of the methanolic extracts was comparable to that of the standard insecticide, deltamethrin (0.2%). The 4% neem leaf extract showed a good controlling effect in comparison with the control but was inferior to the neem seed extracts and deltamethrin. Altogether the neem treatments increased the yield of okra. Especially the 4% methanolic neem seed extract proved to be very efficient, giving even a higher yield than deltamethrin.[45]

Aqueous neem seed extracts (25 and 50 g/l) gave a significant yield increase of okra in Togo due to good control of *S. derogata* and *Bemisia tabaci*.[46]

6. Tomato and Eggplant

An acetonic neem seed kernel extract applied at 2.5% in Mauritius in a field trial reduced the number of larvae of the Afro-Asian bollworm, *Heliothis armigera,* on to-

Table 2
**ATTACK ON EGGPLANT TREATED WITH NEEM KERNEL POWDER
(NKP) AQUEOUS EXTRACT OR NEEM OIL (16 PLANTS/PLOT) BY
JACOBIELLA FACIALIS (FINAL EVALUATION), *SCROBIPALPA
ERGASIMA* AND *PHYCITA MELONGENAE* (AVERAGE OF SIX
EVALUATIONS), AND FEEDING DAMAGE ON LEAVES CAUSED BY
CATERPILLARS AND GRASSHOPPERS (FINAL EVALUATION): SECOND
FIELD TRIAL**

| Treatment | Damage by *J. facialis** | Number of leaves/plant attacked by | | Index of feeding damage** |
		S. ergasima	*P. melongenae*	
Control	33.0 ***	2.1 a	1.4 a	46.9 a
NKP, 25 g/l	8.5 c	0.1 c	0 b	7.5 bc
NKP, 50 g/l	5.8 c	0.1 c	0 b	1.9 c
Neem oil, ULV, 5 1/ha	20.4 b	1.0 b	1.2 a	18.2 b
Neem oil, ULV, 10 1/ha	21.8 b	0.8 b	0.6 a	16.5 b

Note: Plot size 1.80 × 2.70 m, all plots fully randomized.

* Average number of leaves with symptoms (yellow leaf edge)/plant.
** Feeding index calculated as product of average number of leaves damaged per plant and mean damage category 1—4 (1 = <10% leaf surface consumed, 4 = >50% leaf surface consumed).
*** Within columns, means followed by a common letter are not significantly different at $p - 0.01$.

matoes by 50%. However, the incidence of *H. armigera* in the neem plots was still too high to prevent damage. Better results were achieved with a weekly alternate use of the neem extract and deltamethrin (Decis). The combination of neem and the synthetic pyrethroid was even more active than nonalternate use of Decis alone.[15]

A series of field trials in Togo showed that eggplants can be efficiently protected against the leafhopper, *Jacobiella facialis,* the leaf miner, *Scrobipalpa ergasima*, and the leaf-roller, *Phycita melongenae,* by aqueous neem kernel extracts or neem oil (Table 2). At concentrations of 25 and 50 g/l of kernel extracts and 5 and 10 1/ha of neem oil (ULV application), the feeding damage caused by the leafhopper and the lepidopterous larvae was negligible. The efficacy of the aqueous seed kernel extract was significantly better than that of the oil. Treatments with the same neem products could only slightly influence the population dynamics of the carmine spider mite, *Tetranychus cinnabarinus,* and therefore could not prevent damage caused by this pest. However, in comparison to the control the effects of the neem extract were good enough to double or triple the yield.

Eggplants were sprayed or dusted in the U.S. weekly with different neem seed products (ethanolic extract at 2% and dust at 20%) against the flea beetle, *Epitrix fuscula,* and the Colorado potato beetle, *Leptinotarsa decemlineata*. The neem spray was very effective against the insect pests and tripled the yield in comparison to the control plots. There was no significant difference between the treatment with the ethanolic extract and the synthetic insecticide Ammo (cypermethrin).[16]

7. Onion

When an aqueous extract of neem leaves at a concentration of 0.5% was sprayed fortnightly on onions in the field in Mauritius, damage by *Liriomyza trifolii* was considerably diminished. Yields in the neem treatment were four to six times higher than in the control plots.[44]

Table 3
EFFECT OF NEEM KERNEL POWDER (NKP) AQUEOUS
EXTRACT ON *SCROBIPALPA ERGASIMA*, ATTACKING GBOMA

Dosage (g NKP/l)	Leaf harvest (1984)				Plant mortality in plants per plot
	2. V.	25. V.	22. VI.	25. VIII.	25. VIII.
	Mines per plant*				
Control	1.6 a**	3.3 a	2.4 a	8.0 a	11.5 a
12.5	0.1 b	0.3 b	0.4 b	3.9 ab	1.5 b
25	0 b	0.4 b	0.1 b	1.1 b	3.5 b
50	0 b	0.1 b	0 b	0.4 b	2.3 b

Note: Plot size 1.25 × 3.50 m, all plots fully randomized.

*   Means of 40 plants per plot.
**   Within columns, means followed by a common letter are not significantly different
      at $p = 0.01$ (mines) and at $p = 0.05$ (plant mortality).

8. Gboma (*Solanum aethiopicum*)

Gboma, an important vegetable in western Africa, was sprayed in a field trial in Togo with an aqueous extract of ground neem seed kernels obtained with 12.5, 25, and 50 g/l of water. The neem treatments, even at the lowest dosage, reduced the number of mines of the leafminer, *Scrobipalpa ergasima* (Lepidoptera) by more than 80% and increased the yield by 30%. However, the efficacy of this concentration was reduced when the population density increased. The higher concentrations gave good to almost complete protection for several days and doubled, sometimes tripled, the yield in comparison to the control (Table 3).

Grasshoppers, such as *Zonocerus variegatus*, were also deterred from feeding.[46]

9. Potato and Sweet Potato

In laboratory studies in West Germany, larvae of the Colorado potato beetle, *Leptinotarsa decemlineata,* were highly susceptible to aqueous, methanolic, AZT, and MTB/ H$_2$O neem seed kernel extracts.[47] Ethanolic extracts of neem seeds sprayed weekly at 2% on potato plants in the field in the U.S. diminished the number of larvae of the beetle by 80%. Monitor, a standard insecticide, gave no better results.[16]

A formulated 20% AZT-VR-K extract was applied in various concentrations (0.06 to 0.25%) and also an aqueous neem kernel extract (30 g kernels/l) were applied in laboratory and greenhouse trials to potato plants to study the effect of neem on females of *L. decemlineata* in West Germany. When the beetles were exposed for 24 h to leaves sprayed with AZT-VR-K (0.25%) or the aqueous extract and were then offered untreated leaves they laid 86% fewer eggs than in the control. Some females were completely sterile (Figure 1). These results were confirmed by combined field-laboratory experiments indicating that it may be possible to control *L. decemlineata* by spraying potato fields with neem seed extracts after migration of the adults in spring when the egg-laying period is starting.[48] In field studies with methanolic extracts and neem oil the efficacy of neem extracts against *L. decemlineata* could be increased by adding different solvents, adjuvants, and emulsifiers to the pure extracts.[7,12,48]

In India, a field experiment was carried out to control the tortoise beetle, *Chirida bipunctata,* on sweet potato. In the different treatments an acetone extract of neem leaves and seeds (25 g plant material per 100 ml acetone) at a dosage of 1%, and neem oil (0.5%) were applied to sweet potato plots at 10-d intervals starting 30 d after

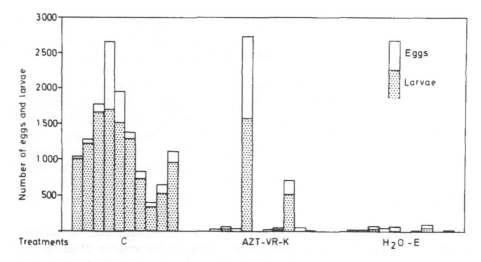

FIGURE 1. Fecundity of ten females of *Leptinotarsa decemlineata* and egg-fertility after uptake of neem seed kernel extracts. AZT-VR-K = enriched, formulated extract; H$_2$O-E = aqueous extract.[48]

transplantation of the vines. Furthermore, two organotins (fentin acetate, fentin hydroxide) were tested. The lowest number of feeding holes per leaf were obtained by the neem seed kernel extract, neem oil, and fentin hydroxide, followed by neem leaf extract and fentin acetate. However, the neem oil negatively affected the development of the plant so that the yields were considerably higher than in the control but lower than in the neem seed and fentin hydroxide plots.[20]

B. FRUIT TREES
1. Citrus

In laboratory trials in the U.S., it was found that adult females of the citrus red mite, *Panonychus citri*, were repelled by leaves treated with neem kernel hexane extracts (0.1%) and laid eggs only on untreated leaves.[49] Good controlling effects with neem oil (1.4%) against *P. citri* were also obtained in laboratory and field trials in the People's Republic of China.[17]

When placed together on bean leaves treated with different neem seed kernel extracts in the laboratory in Israel, the carmine spider mite, *Tetranychus cinnabarinus*, was 3 to 58 times (depending on the solvent) more susceptible to the neem treatment than the predacious mite, *Phytoseiulus persimilis*. The order of toxicity of the extracts against *T. cinnabarinus* was pentane > acetone > ethanol > methanol. All extracts were nontoxic to the predatory spider, *Chiracanthium mildei*, at 2.5%.[50]

A hexane extract of neem seeds proved to be efficient against the California red scale, *Aonidiella aurantii*, the yellow scale, *A. citrina* and the citrus mealybug, *Planococcus citri*. In laboratory tests in the U.S., concentrations of 1% hexane extract reduced the number of scale insects on lemon fruits or seedlings by 92% (red scale), 90% (yellow scale), and 46% (citrus mealybug), respectively. In further experiments fractions of the hexane extract applied to lemon seedlings reduced the numbers of the woolly whitefly, *Aleurothrixus floccosus*.[49]

In laboratory and field trials in China, the application of 1.4% emulsified neem seed oil was effective as a repellent and feeding deterrent against the Asiatic citrus psyllid, *Diaphorina citri*.[17]

In a field test in India, the application of neem cake in water suspension (1 kg neem cake per 10 l) efficiently protected young citrus seedlings against the citrus leafminer,

Phyllocnistis citrella. The neem treatment proved to be effective up to 14 d and gave better results than most of the tested synthetic insecticides applied at the same time.[51] Neem oil at 1.4% also proved to be effective against the leafminer in field and laboratory tests in China.[17]

To protect sweet orange (*Citrus sinensis*) seedlings from attack by larvae of the swallowtail butterfly, *Papilio demodocus*, the plants were treated in the nursery in Gambia once a week with an aqueous extract of ripe neem fruits ("berries") at a concentration of 112.5 g/l. In comparison to the control, the number of seedlings damaged was nearly the same in the neem treatment but the degree of damage was considerably reduced.[21]

A methanolic neem seed kernel extract at 1% was applied to banana fruits in the laboratory to deter oviposition of the oriental fruit fly, *Dacus dorsalis*. The extract completely inhibited egg-laying of the fruit fly for only 2 d. The number of eggs in the fruits then increased rapidly.[52]

It could be demonstrated in laboratory trials in West Germany that methanolic neem seed extracts, incorporated in an artificial diet, caused disturbance of larval growth of *Ceratitis capitata* and increased mortality. The extracts also affected the fitness of resulting adult fruit flies even at low concentrations. However, under field conditions a spraying of neem products on citrus fruits seems to be not very promising for control of fruit flies because oviposition is not sufficiently deterred and larvae will not come in contact with neem substances after emergence.[53]

2. Other Fruit Trees

In laboratory tests in Israel, avocado leaves were dipped in aqueous neem seed "suspensions" of various concentrations and fed to larvae of the giant looper, *Boarmia* (*Ascotis*) *selenaria*. A 0.3% neem seed extract caused 80% mortality until the 18th d of the experiment. At lower concentrations of 0.1 and 0.05% the larval growth was strongly inhibited.[54]

Larvae of the gypsy moth, *Lymantria dispar*, which is a polyphagous insect attacking many forest trees and fruit trees, proved to be highly susceptible to the active substances of neem. Neem seed kernel powder, incorporated in an artificial diet to a final concentration of 0.02%, caused complete mortality of second instar larvae after 30 d. Higher concentrations accelerated drastically the mortality of the larvae. In further tests in West Germany, an aqueous extract of neem seed kernels at 0.5%, sprayed on oak leaves, killed all larvae within 15 d. It could be demonstrated in field tests that the residues of the aqueous extract were still active after 12 d.[55]

IV. EXPERIMENTAL RESULTS WITH NEMATODES ON VEGETABLES

The use of neem seed cake and neem leaves, incorporated into the soil, reduced the incidence of root knots caused by nematodes of the genus *Meloidogyne* and improved the development of tomatoes and okra plants in India.[56,57] Similar results were obtained with eggplant.[58]

Aqueous extracts of neem seed cake inhibited the penetration of nematodes into the roots of various crops.[59,60]

In laboratory trials in India, the toxicities of aqueous extracts of neem leaf, flower, fruit, bark, root, and gum were compared with one another against the nematodes *Helicotylenchus indicus*, *Hoplolaimus indicus*, *Tylenchus filiformis*, *Tylenchorhynchus brassicae*, *Rotylenchus reniformis*, and *Meloidogyne incognita*. All plant parts possessed nematocidal properties. In general, the order of effectiveness was fruit, leaf, bark and flower, gum, and root.[61,62]

Further studies in South Asia confirmed that neem cake especially contains substances such as nimbidin and thionemone which are highly active against various nematodes.[63]

In further studies, nimbin, salannin, deacetylnimbin, 4-epinimbin, nimbinene, deacetylnimbinene, and certain novel as well as unidentified compounds were treated with carbofuran in comparison. It was demonstrated that deacetylnimbin and certain fractions (NP-1, P-1, and P-2) produced the strongest effects.[64] When chickpea seeds were soaked for 3 and 6 h in different concentrations of aqueous extracts of neem seed cake the damage by *M. incognita* was significantly reduced. Shorter soaking periods and lower concentrations of the extract did not affect chickpea seed germination.[66]

Neem seed cake was applied to the soil at a dosage of 1700 kg/ha in comparison with aldicarb granules (4 kg a.i./ha) or fensulfothion (1% w/w) seed treatment to protect *Vigna radiata* plants against *Tylenchorhynchus* sp., *Helicotylenchus* sp., *R. reniformis*, and *M. incognita* in India. The neem treatment increased grain yield by 42% and was considerably better than the aldicarb treatment, increasing the yield by 8.9%. However, the best result (63% yield increase) was achieved by the synthetic nematocide fensulfothion.[66]

Ground neem leaves and seed kernels were mixed in West Germany with soil at a rate of 1% (v:v). The mixtures favored development of tomato plants by reducing the number of root balls caused by *M. arenaria*. Thereby, seed kernels were more efficient than neem leaves. The neem treatment reduced significantly the number of *Pratylenchus penetrans* extracted from the roots. A field trial confirmed these findings. Although there was no significant difference in the numbers of some nematode species in the treated and control plots, the majority of the nematodes was extracted from the roots of plants in untreated soil.[67]

V. CONCLUSIONS

According to results of numerous field experiments in Africa, America, Asia, and Europe, neem products can control effectively a considerable number of vegetable and some fruit-tree pests. The best results were obtained, as a rule, by application of extracts from neem seeds with alcohols (methanol, ethanol). Extracts from neem leaves were generally less effective or, in other words, greater amounts of raw material had to be used to obtain results comparable to those of seed extracts. Neem oil also showed a lower degree of efficacy than seed extracts. However, the effect of neem products varied with the pest species tested and the progeny of the materials used which is not surprising due to problems of standardization. Therefore exact comparisons are not possible.

Lepidopterous larvae were the main target pests of neem application.[68] They are the most important group of harmful insects in warm climates. In most field experiments against caterpillars satisfactory to very good results were obtained, especially by application of seed kernel extracts. The feeding activity and metamorphosis (molts) of the insects were seriously disturbed by the active ingredient(s) of these extracts, leading to a decline of pest populations which was often at par with the effect of a number of well-known synthetic pesticides. In some cases the effects of neem products were better than those of synthetic products, especially in control of the diamondback moth, *Plutella xylostella*. The latter develops resistance to all major groups of pesticides rather quickly, especially in Asia, but also to microbial pesticides based on *Bacillus thuringiensis*.[69] Interestingly, a selection for resistance of *P. xylostella* against neem products during 40 generations in the laboratory was not successful.[70]

Other insects which showed sensitivity to neem products were freely feeding larvae of Coleoptera, for instance of phytophagous Coccinellidae. The same applied to some polyphagous grasshoppers. Adult beetles were deterred from feeding on treated plants, for instance *Popillia japonica, Ootheca bennigseni,* and various flea beetles (*Phyllotreta, Podagrica*).

Adult beetles may be sterilized by neem products. This has been demonstrated in the case of the Colorado potato beetle, *Leptinotarsa decemlineata.*[48] Sterilizing effects are also observed after uptake of low concentrations of neem seed extracts by larvae of some Lepidoptera and Diptera.

Another group of insects which proved to be susceptible to neem products are mining flies (Agromyzidae). These serious pests of vegetables and ornamentals, which may also develop resistance to synthetic pesticides rather quickly, were controlled by a number of neem seed extracts and reacted even after watering of host plants, which means that a systemic effect was obtained.[71]

Whereas leafhoppers, mainly *Jacobiella facialis,* were well controlled by application of seed kernel water extracts, some other homopterans, such as aphids, are in general less susceptible to neem products with a low content of oil. Therefore, oily formulations may be more promising against this group of insects.

Hexane extracts of neem seeds were effective against scale insects and mealybugs on citrus fruits and seedlings, and to some extent also against whiteflies. Pentane extracts gave the best results against spider mites and proved to be considerably less toxic to predaceous mites and spiders.[50] More field experiments are needed with hexane and pentane extracts to get a clear picture of the potential of such products, especially for control of spider mites and homopterous insects on fruit trees but also for combating mites on a number of vegetables.

In some cases neem products could not control the target insects on vegetables, for instance fruitflies of the genus *Dacus* on cucurbits. This largely depended on the egg-laying behavior of these flies.

Some nematode species, especially *Meloidogyne* spp., were regularly reduced by incorporation of neem seed cake or neem leaves in the soil but considerable amounts were needed to obtain this desirable effect. However, such a practice has been followed for many years by cardamom farmers in parts of India, apparently with good success.[72]

There are several active principles in neem seeds, but the dominating substance is the tetranortriterpenoid, azadirachtin. Normally the best results in controlling insect pests are obtained with seed extracts containing the highest amounts of this compound.[73] For this reason it is of great importance to collect seeds of good quality and to dry and store them properly to prevent losses of active ingredient and contamination by fungi causing rotting in the store, especially under humid conditions.

Neem leaves seem to be free of azadirachtin but contain a number of related compounds with a similar mode of action.

The components in neem cake, which show activity against nematodes, are nimbin, deacetylnimbin, and thionemone.

Aqueous and alcoholic neem extracts have to be applied carefully as high volume sprays to protect the plants to be treated well. In case of pests attacking the lower surface of the leaves, both leaf surfaces have to be sprayed. Neem oil may be applied with water as high volume spray (low concentrations) or as low to ultra low volume spray (high concentration of oil). By using these products there might be some danger of phytotoxicity.

Rainfall is destructive to most neem products because they are washed down from the plants. On the other hand, dry and warm weather improves the effect of neem applications, partly by increasing the feeding activity of the target insects. Exact timing of neem applications may also play an important role.

In developing countries the use of neem seed water extracts, neem oil, and neem cake against vegetable and fruit-tree pests is feasible even at the peasant farmer's level. On the other hand, in industrialized countries and India some alcoholic, formulated neem seed extracts have been/are being developed. Although their efficacy is sometimes inferior to that of toxic, broad-spectrum, synthetic insecticides, the degree of reduction of pest populations is, as a rule, quite sufficient for instance in home gardening or under similar conditions where the quality of the products does not play an outstanding role.[73]

The lack of toxicity of neem seed products to warm-blooded animals and their relative selectivity concerning natural enemies of pests are outstanding, desirable criteria which may lead to an increasing popularity of these natural pesticides in the future in spite of their delayed effect. These features make them very suitable insecticides for insect pest management (IPM) programs especially in vegetables.

REFERENCES

1. **Dreyer, M.,** Field and laboratory trials with simple neem products as protectants against pests of vegetable and field crops in Togo, in *Proc. 3rd Int. Neem Conf.,* Nairobi, Kenya, 1986, Schmutterer, H. and Ascher, K. R. S., Eds., GTZ Press, Eschborn, West Germany, 1987, 431.
2. **Adhikary, S.,** Results of field trials to control the diamondback moth, *Plutella xylostella* L., by application of crude methanolic extracts and aqueous suspensions of seed kernels and leaves of neem, *Azadirachta indica* A. Juss, in Togo, *Z. Angew. Entomol.,* 100, 27, 1985.
3. **Steets, R.,** The effect of crude extracts from the meliaceous plants *Azadirachta indica* and *Melia azedarach* on some insect species, *Z. Angew. Entomol.,* 77, 306, 1976.
4. **Karel, A. K.,** Response of *Ootheca bennigseni* (Coleoptera: Chrysomelidae) to neem extracts, in *Proc. 3rd Int. Neem Conf.,* Nairobi, Kenya, 1986, Schmutterer, H. and Ascher, K. R. S., Eds., GTZ Press, Eschborn, West Germany, 1987, 393.
5. **Larson, R. O.,** Development of Margosan-O, a pesticide from neem seed, *Proc. 3rd Int. Neem Conf.,* Nairobi, Kenya, 1986, Schmutterer, H. and Ascher, K. R. S., Eds., GTZ Press, Eschborn, West Germany, 1987, 243.
6. **Feuerhake, K.,** Effectiveness and selectivity of technical solvents for the extraction of neem seed components with insecticidal activity, in *Natural Pesticides from the Neem Tree and Other Tropical Plants,* Schmutterer, H. and Ascher, K. R. S., Eds., GTZ Press, West Germany, 1984, 103.
7. **Feuerhake, K. and Schmutterer, H.,** Development of a crude standardized and formulated insecticide from crude neem kernel extract, *Z. Pflanzenkr. Pflanzenschutz,* 92, 643, 1985.
8. **Khan, M. W., Alam, M. M., Khan, A. M., and Saxena, S. K.,** Effect of water soluble fractions of oil-cakes and bitter principles of neem on some fungi and nematodes, *Acta Bot. Indica,* 2, 120, 1974.
9. **Sombatsiri, K. and Temboonkat, K.,** Efficacy of an improved neem kernel extract in the control of *Spodoptera litura* and of *Plutella xylostella* under laboratory conditions and in field trials, *Proc. 3rd Int. Neem Conf.,* Nairobi, Kenya, 1986, Schmutterer, H. and Ascher, K. R. S., Eds., GTZ Press, Eschborn, West Germany, 1987, 195.
10. **Kirsch, K.,** Studies on the efficacy of neem extracts in controlling major pests in tobacco and cabbage, *Proc. 3rd Int. Neem Conf.,* Nairobi, Kenya, 1986, Schmutterer, H. and Ascher, K. R. S., Eds., GTZ Press, Eschborn, West Germany, 1987, 495.
11. **Lange, W. and Feuerhake, K.,** Improved activity of enriched neem seed extracts with synergists (piperonyl butoxide) under laboratory conditions, *Z. Angew. Entomol.,* 98, 368, 1984.
12. **Lange, W. and Schmutterer, H.,** Attempts to improve the growth-disrupting activity of a crude methanolic extract of neem seeds (*Azadirachta indica*) with synergists, *Z. Pflanzenkr. Pflanzenschutz,* 89, 258, 1982.
13. **Fagoonee, I.,** Behavioral response of *Crocidolomia binotalis* to neem, in *Natural Pesticides from the Neem Tree (Azadirachta indica A. Juss),* Schmutterer, H., Ascher, K. R. S., and Rembold, H., Eds., GTZ Press, Eschborn, West Germany, 1981, 109.

14. **Fagoonee, I.,** The potential of natural products in crop protection in Mauritius, in *Proc. Neth. Agric. Prod. Conf.,* 1979, Reduit, Mauritius, 1980, 201.

15. **Fagoonee, I.,** Use of neem in vegetable crop protection in Mauritius, in *Proc. 3rd Int. Neem Conf.,* Nairobi, Kenya, 1986, Schmutterer, H. and Ascher, K. R. S., Eds., GTZ Press, Eschborn, West Germany, 1987, 419.

16. **Reed, D. K. and Reed, G. L.,** Control of vegetable insects with neem seed extracts, in *Proc. Indiana Acad. Sci.,* 94, 335, 1986.

17. **Chiu, Shin-Foon,** The active principles and insecticidal properties of some Chinese plants, with special reference to Meliaceae, in *Natural Pesticides from the Neem Tree and Other Tropical Plants,* Schmutterer, H. and Ascher, K. R. S., Eds., GTZ Press, Eschborn, West Germany, 1984, 255.

18. **Kirsch, K.,** Effect of Neem Seed Extracts on Insect Pests of Tobacco, Cotton and Vegetables in the Philippines, Ph.D. thesis, University of Giessen, West Germany, 1987.

19. **Lange, W.,** Experiments on Synergistic and Residual Effects of Neem Seed Extracts, Ph.D. thesis, University of Giessen, West Germany, 1985.

20. **Abdul Kareem, A.,** Neem as an antifeedant for certain phytophagous insects and a bruchid on pulses, in *Natural Pesticides from the Neem Tree (Azadirachta indica A. Juss),* Schmutterer, H., Ascher, K. R. S., and Rembold, H., Eds., GTZ Press, Eschborn, West Germany, 1981, 223.

21. **Redknap, R. S.,** The use of crushed neem berries in the control of some insect pests in Gambia, in *Natural Pesticides from the Neem Tree (Azadirachta indica A. Juss),* Schmutterer, H., Ascher, K. R. S., and Rembold, H., Eds., GTZ Press, Eschborn, West Germany, 1981, 205.

22. **Dreyer, M.,** Effects of aqueous neem extracts and neem oil on the main pests of *Cucurbita pepo* in Togo, in *Natural Pesticides from the Neem Tree and Other Tropical Plants,* Schmutterer, H. and Ascher, K. R. S., Eds., GTZ Press, Eschborn, West Germany, 1984, 435.

23. **Reed, D. K., Warthen, J. D., Jr., Ubel, E. C., and Reed, G. L.,** Effects of two triterpenoids from neem on feeding by cucumber beetles (Coleoptera: Chrysomelidae), *J. Econ. Entomol.,* 75, 1109, 1982.

24. **Coudriet, D. L., Probhaker, N., and Meyerdirk, D. E.,** Sweetpotato whitefly (Homoptera:Aleyrodidae): effects of neem seed extract on oviposition on immature stages, *Environ. Entomol.,* 14, 776, 1985.

25. **Singh, R. P. and Srivastava, B. G.,** Alcohol extract of neem (*Azadirachta indica* A. Juss) seed oil as oviposition deterrent for *Dacus*

cucurbitae (Coq.), *Indian J. Entomol.,* 45, 497, 1983.

26. **Mansour, F. A. and Ascher, K. R. S.,** Effects of neem (*Azadirachta indica*) seed kernel extracts from different solvents on the carmine spider mite, *Tetranychus urticae,* in *Natural Pesticides from the Neem Tree and Other Tropical Plants,* Schmutterer, H. and Ascher, K. R. S., Eds., GTZ Press, Eschborn, West Germany, 1984, 461.

27. **Hongo, H. and Karel, A. K.,** Effect of plant extracts on insect pests of common beans, *J. Appl. Entomol.,* 102, 164, 1986.

28. **Schauer, M.,** Effects of variously formulated neem seed extracts on *Acyrthosiphon pisum* and *Aphis fabae,* in *Natural Pesticides from the Neem Tree and Other Tropical Plants,* Schmutterer, H. and Ascher, K. R. S., Eds., GTZ Press, Eschborn, West Germany, 1984, 141.

29. **Steets, R. and Schmutterer, H.,** Influence of azadirachtin on longevity and reproduction capacity of *Epilachna varivestis* Muls. (Coleoptera:Coccinellidae), *Z. Pflanzenkr. Pflanzenschutz,* 82, 176, 1975.

30. **Schmutterer, H. and Rembold, H.,** On the effect of some pure fractions from seeds of *Azadirachta indica* on feeding activity and metamorphosis of *Epilachna varivestis* (Col., Coccinellidae), *Z. Angew. Entomol.,* 89, 179, 1980.

31. **Ascher, K. R. S. and Gsell, R.,** The effect of neem seed kernel extract on *Epilachna varivestis* Muls. larvae, *Z. Pflanzenkr. Pflanzenschutz,* 88, 764, 1981.

32. **Ladd, T. L.,** Neem seed extracts as feeding deterrents for the Japanese beetle, *Popillia japonica,* in *Natural Pesticides from the Neem Tree (Azadirachta indica A. Juss),* Schmutterer, H., Ascher, K. R. S., and Rembold, H., Eds. GTZ Press, Eschborn, West Germany, 1981, 149.

33. **Adhikary, S.,** The Togo experience in moving from neem research to its practical application for plant protection, in *Natural Pesticides from the Neem Tree (Azadirachta indica A. Juss),* Schmutterer, H., Ascher, K. R. S., and Rembold, H., Eds., GTZ Press, Eschborn, West Germany, 1981, 215.

34. **Gohokar, R. T., Thakre, S. M., and Borle, M. N.,** Chemical control of gram pod borer (*Heliothis armigera* H.) by different synthetic pyrethroids and insecticides, *Pesticides,* 55, 1987.

35. **Jain, H. K., Srivastava, K. P., Agnihotri, N. P., and Cajbhiye, V. T.,** Evaluation of bioefficacy of some insecticides against *Heliothis armigera* and *Melanagromyza obtusa* and their residues on pigeon pea, *Int. Pigeonpea Newslett.,* 1986.

36. **Srivastava, K. P., Agnihotri, N. P., Gajbhiye,**

V. T., and Jain, H. K., Relative efficacy of fenvalerate, quinalphos and neem kernel extracts for the control of pod fly, *Melanagromyza obtusa* (Malloch) and pod borer, *Heliothis armigera* (Hubner) infesting red gram, *Cajanus cajan* (L.) Millsp. together with their residues, *J. Entomol. Res.*, 8, 1, 1984.

37. **Parmar, B. S.,** An overview of neem research and use in India during the years 1983—1986, in *Proc. 3rd Int. Neem Conf.*, Nairobi, Kenya, 1986, Schmutterer, H. and Ascher, K. R. S., Eds., GTZ Press, Eschborn, West Germany, 1987, 55.

38. **Parmar, B. S. and Srivastava, K. P.,** Development of some neem formulations and their evaluation for the control of *Spilosoma obliqua* (Wlk.) in the laboratory and *Euchrysops cnejus* (F.) in the field, in *Proc. 3rd Int. Neem Conf.*, Nairobi, Kenya, 1986, Schmutterer, H. and Ascher, K. R. S., Eds., GTZ Press, Eschborn, West Germany, 1987, 205.

39. **Patel, H. K., Patel, V. C., Chari, M. S., Patel, J. C., and Patel, J. R.** Neem seed paste suspension—a sure deterrent to hairy caterpillar (*Amsacta moorei* Butl.), *Madras Agric. J.*, 55, 509, 1968.

40. **Saxena, R. C.,** Note on the use of neem kernel for the protection of dew-gram against *Amsacta moorei* Butler, *Indian J. Agric. Sci.*, 52, 51, 1982.

41. **Larew, H. G., Webb, R. E., and Warthen, J. D., Jr.,** Leafminer controlled on chrysanthemum by neem seed extract applied to potting soil, in *Proc. 4th Annu. Ind. Conf. on Leafminers,* Sarasota, Fl., 1984, 108.

42. **Webb, R. E., Larew, H. G., Wieber, A. M., Ford, P. W., and Warthen, J. D.,** Jr., Systemic activity of neem seed extract and purified azadirachtin against *Liriomyza* leafminers, in *Proc. 4th Annu. Ind. Conf. on Leafminers,* Sarasota, Fl., 1984, 118.

43. **Webb, R. E., Hinebaugh, M. A., Lindquist, R. K., and Jacobson, M.,** Evaluation of aqueous solution of neem seed extract against *Liriomyza sativae* and *L. trifolii* (Diptera:Agromyzidae), *J. Econ. Entomol.*, 76, 357, 1983.

44. **Fagoonee, I. and Toory, V.,** Contribution to the study of the biology and ecology of the leafminer, *Liriomyza trifolii*, and its control by neem, *Insect Sci. Appl.*, 5, 23, 1984.

45. **Adhikary, S.,** Results of field trials to control common insect pests of okra, *Hibiscus esculentus* L., in Togo by application of methanolic extracts of leaves and seed kernels of the neem tree, *Azadirachta indica* A. Juss, *Z. Angew. Entomol.*, 98, 327, 1984.

46. **Dreyer, M.,** Investigations on the Effect of Aqueous Extracts and Other Products from Neem Seeds Against Pests of Vegetable and Field Crops in Togo, Ph.D. thesis, University of Giessen, West Germany, 1986.

47. **Feuerhake, K.,** Investigations on the Production and Formulation of Seed Ingredients of the Neem Tree (*Azadirachta indica* A. Juss) with Regard to Their Use as Pesticides in Developing Countries, Ph.D. thesis, University of Giessen, West Germany, 1985.

48. **Schmutterer, H.,** Fecundity-reducing and sterilizing effects of neem seed kernel extracts in the Colorado potato beetles, *Leptinotarsa decemlineata*, in *Proc. 3rd Int. Neem Conf.,* Nairobi, Kenya, 1986, Schmutterer, H. and Ascher, K. R. S., Eds., GTZ Press, Eschborn, West Germany, 1987, 351.

49. **Jacobson, M., Reed, D. K., Crystal, M. M., Moreno, D. S., and Soderstrom, E. L.,** Chemistry and biological activity of insect feeding deterrents from certain weed and crop plants, *Entomol. Exp. Appl.*, 24, 448, 1978.

50. **Mansour, F., Ascher, K. R. S., and Omari, N.,** Effect of neem seed kernel extracts from different solvents on the predacious mite *Phytoseiulus persimilis* and the phytophagous mite *Tetranychus cinnabarinus* as well as the predatory spider, *Chiracanthium mildei*, in *Proc. 3rd Int. Neem Conf.*, Nairobi, Kenya, 1986, Schmutterer, H. and Ascher, K. R. S., Eds., GTZ Press, Eschborn, West Germany, 1987, 577.

51. **Batra, R. C. and Sandhu, G. S.,** Comparison of different insecticides for the control of citrus leaf-miner in the nursery, *Pesticides,* 15, 5, 1981.

52. **Sombatsiri, K. and Tigvattanont, S.,** Effects of neem extract on some insect pests of economic importance in Thailand, in *Natural Pesticides from the Neem Tree and Other Tropical Plants,* Schmutterer, H. and Ascher, K. R. S., Eds., GTZ Press, Eschborn, West Germany, 1984, 95.

53. **Steffens, R. J. and Schmutterer, H.,** The effect of crude methanolic neem (*Azadirachta indica*) seed kernel extract on metamorphosis and quality of adults of the Mediterranean fruit fly, *Ceratitis capitata* Wied. (Diptera, Tephritidae), *Z. Angew. Entomol.*, 94, 98, 1982.

54. **Meisner, J., Wysoki, M., and Ascher, K. R. S.,** The residual effect of some products from neem (*Azadirachta indica* A. Juss) seeds upon larvae of *Boarmia (Ascotis) selenaria* Schiff. in laboratory trials, *Phytoparasitica*, 4, 185, 1976.

55. **Skatulla, U. and Meisner, J.,** Laboratory experiments using neem seed extract for control of the gypsy moth, *Lymantria dispar* L., *Anz. Schaedlingskd. Pflanz. Umweltschutz,* 48, 38, 1975.

56. **Singh, R. S. and Sitaramaiah, K.,** Incidence of root-knot of okra and tomatoes in oil-cake amended soil, *Plant Dis. Rep.*, 50, 668, 1966.

57. **Singh, R. S. and Sitaramaiah, K.,** Effect of decomposing green leaves, saw-dust and urea on the incidence of root-knot of okra and tomato, *Indian Phytopathol.,* 20, 349, 1967.

58. **Lall, B. S. and Hameed, S. F.,** Studies on the biology of the root-knot nematodes (*Meloidogyne* spp.) with special reference to host resistance and manuring, in *Proc. 1st All-India Nematol. Symp.,* New Delhi, India, 1969.

59. **Khan, W. M., Khan, A. M., and Saxena, S. K.,** Influence of certain oil-cake amendments on nematodes and fungi in tomato field, *Acta Bot. Ind.,* 1, 49, 1973.

60. **Gill, J. S. and Lewis, G. T.,** Systemic action of an insect feeding deterrent, *Nature,* 232, 402, 1971.

61. **Siddiqui, M. A. and Alam, M. M.,** Evaluation of nematocidal properties of different parts of Margosa and Persian lilac, *Neem Newslett.,* 1, 1, 1985.

62. **Siddiqui, M. A. and Alam, M. M.,** Further studies on the nematode toxicity of Margosa and Persian lilac, *Neem Newslett.,* 4, 43, 1985.

63. **Khan, W. M., Alam, M. M., and Rais, A.,** Mechanism of control of plant parasitic nematodes as a result of the application of oil-cakes to the soil, *Indian J. Nematol.,* 4, 93, 1974.

64. **Devakumar, C., Goswami, B. K., and Mukerjee, S. K.,** Nematicidal principles from neem (*Azadirachta indica* A. Juss). I. Screening of neem kernel fractions against *Meloidogyne incognita* (Kofoid & White) Chitwood, *Indian J. Nematol.,* 15, 121, 1985.

65. **Devakumar, C., Mukerjee, S. K., and Goswami, B. K.,** Chemistry and nematocidal activity of neem (*Azadirachta indica*) constituents, *Abstr. Natl. Symp. Insecticidal Plants and Control of Environ. Pollution,* Bharathidasan University, Tiruchirapalli, India, 1986, 18.

66. **Vijayalakshmi, K., Gaur, H. S., and Mishra, S. D.,** Effect of soaking seed in extracts of oilseed cakes and plant leaves on the germination of chickpea and its vulnerability to the root-knot nematode, *Abstr. Natl. Symp. On Soil Pests and Soil Organisms,* Banaras Hindu University, Varanasi, India, 1984, 71.

67. **Mishra, S. D. and Gaur, H. S.,** Control of nematodes infesting mung with nematicidal seed treatment and field applications of dasanit, aldicarb and neem cake, *Abstr. Natl. Symp. on Soil Pests and Soil Organisms,* Banaras Hindu University, Varanasi, India, 1984, 71.

68. **Rossner, J. and Zebitz, C. P. W.,** Effect of neem products on nematodes and growth of tomato (*Lycopersicon esculentum*) plants, in *Proc. 3rd Int. Neem Conf.,* Nairobi, Kenya, 1986, Schmutterer, H. and Ascher, K. R. S., Eds., GTZ Press, Eschborn, West Germany, 1987, 611.

69. **Schmutterer, H.,** Which insect pests can be controlled by application of neem seed kernel extracts under field conditions?, *Z. Angew. Entomol.,* 100, 468, 1985.

70. **Kirsch, K. and Schmutterer, H.,** Low efficacy of a *Bacillus thuringiensis* (Berl.) formulation in controlling the diamondback moth, *Plutella xylostella* (L.), in the Philippines, *Z. Angew. Entomol.,* 103, in press.

71. **Vollinger, M.,** The possible development of resistance against neem seed kernel extracts and deltamethrin in *Plutella xylostella,* in *Proc. 3rd Int. Neem Conf.,* Nairobi, Kenya, 1986, Schmutterer, H. and Ascher, K. R. S., Eds., GTZ Press, Eschborn, West Germany, 1987, 543.

72. **Ahmed, S. and Koppel, B.,** Use of neem and other botanical material for pest control by farmers in India: summary of findings, in *Proc. 3rd Int. Neem Conf.,* Nairobi, Kenya, 1986, Schmutterer, H. and Ascher, K. R. S., Eds., GTZ Press, Eschborn, West Germany, 1987, 623.

73. **Ermel, K., Pahlich, E., and Schmutterer, H.,** Azadirachtin content of neem kernels from different geographical locations, and its dependence on temperature, relative humidity, and light, in *Proc. 3rd Int. Neem Conf.,* Nairobi, Kenya, 1986, Schmutterer, H. and Ascher, K. R. S., Eds., GTZ Press, Eschborn, West Germany, 1987, 171.

5 Effect of Neem on Insect Pests of Ornamental Crops Including Trees, Shrubs, and Flowers

Hiram G. Larew
U.S. Department of Agriculture
Agricultural Research Service
Florist and Nursery Crops Laboratory
Beltsville, Maryland

TABLE OF CONTENTS

I. INTRODUCTION

Although considered by some as insignificant, in reality, the production and maintenance of ornamental crops is a multibillion dollar per year industry involving trade of several million metric tons between underdeveloped, developing, and developed countries.[1-3] Several Central American and African countries rely on the production of cut flowers as one of their largest commodity exports. Compared to most agronomic crops, ornamentals are considered speciality, high-value crops,[4] and the cost of protecting them from pests is similarly steep. In Florida, for example, the cost of treating chrysanthemums for suppression of the agromyzid leafmining fly, *Liriomyza trifolii* (Burgess), was estimated at between $360 to 560/ha/year ($900 to 1400/acre/year).[5] One reason that costs are high is that with few exceptions, the insects that attack ornamentals have become increasingly resistant to synthetic insecticides[6,7] and growers often respond by applying more insecticide more often to minimize cosmetic damage to an aesthetic crop. Additionally, because of safety concerns, fewer materials are available for growers to use, and those that are have become increasingly expensive. As an alternative, neem seed extract (NSE) offers promise.

Until recently, efficacy of NSE as an insecticide had been demonstrated primarily against pests of food and fiber crops. Fewer accounts are available of the NSE activity against pests of ornamentals. For example, of the 85 species of insects listed by Warthen[8] against which NSE has been used, only 6 attack ornamentals, and of the 26 insect species listed by Ahmed et al.[9] as being controlled by NSE, none is a pest of ornamentals. An updated tally of research, however, would show a change. More pests of ornamentals have recently been studied, and results have focused attention on the usefulness of neem in developed countries. Studies with one insect, *L. trifolii*, provide one of the more comprehensive sets of studies available on NSE (see Section III.E.1). The fact that Margosan-O, the first commercially available NSE-insecticide in the U.S., has been registered for use against pests on nonfood crops highlights the potential importance of neem in the ornamentals industry. The purpose of this chapter is to review the published information on NSE and Margosan-O when used against insects on ornamentals. Literature published before June 1987 is reviewed.

II. TREE PESTS

A. GYPSY MOTH (*LYMANTRIA DISPAR* [L.])

The gypsy moth, *L. dispar* (L.) (Lepidoptera:Liparidae), is one of the most destructive pests of forest and shade trees in the U.S.[10] Two separate trials demonstrated that gypsy moth larvae are sensitive to aqueous extracts of neem seeds when the extracts are incorporated into artifical diet or sprayed on oak (*Quercus*) leaves, a preferred food of the larvae.[11,12] Larval development was slowed and molting was prevented by NSE. Mortality was 100% at 0.02, 0.2, and 0.5% concentrations. Residual activity was observed 12 d after spraying leaves of *Q. robur* with NSE. In another study, although the incidence of larval mortality due to virus infection was high, NSE nonetheless caused appreciable larval mortality when fourth and fifth instars were fed oak leaves sprayed with 0.5% NSE.[13]

B. SAWFLY (*PRISTOPHORA ABIETINA* [CHRIST])

P. abietina (Christ) (Hymenoptera:Tenthredinidae) is a serious sawfly pest of young spruce trees in central Europe. Larvae feed on young needles. "Enriched" formulations[14] of NSE were applied to spruce branches as damage began to appear when larvae were first and second instars. Larval populations were reduced by 50% 6 d after treatment.[15] "Only a few ill-looking larvae were left" 10 d after treatment. Feeding damage on treated branches was slight compared to untreated ones.

C. BIRCH LEAFMINER (*FENUSA PUSILLA* [LEPELETIER])

The birch leafminer, *F. pusilla* (Lepeletier) (Hymenoptera:Tenthredinidae), damages foliage of birches, *Betula* spp., by mining through leaves as larvae. When applied to branches as a foliar spray, 1% NSE caused as much leafminer mortality as did comparable sprays of Metasystox-R[16,17] NSE did not prevent mine development on foliage (Metasystox-R did), but caused nearly 100% mortality in the 93 branches that were sprayed and sampled. Thus NSE was as thorough, but not as quick, in killing larval birch leafminers. Time of spray did not affect efficacy; sprays at oviposition and at early instar gave similar results. Thus NSE sprays might greatly reduce local populations of birch leafminers in generations subsequent to treatment. Whole-tree and area-wide (e.g., nursery, lawn) treatments, however, are needed to realistically assess the use of NSE against the birch leafminer.

D. ORANGE STRIPED OAKWORM (*ANISOTA SENATORIA* [J. E. Smith])

When 0.2% NSE was sprayed on branches of willow oak (*Q. phellos* L.) infested with larval orange striped oakworms (*A. senatoria* [J. E. Smith]) (Lepidoptera:Citheroniidae), no larval mortality was observed 24 and 96 h after treatment.[18]

E. SUMMARY

Results from trials against insect pests of trees indicate that NSE is effective as a spray under small trial conditions against three of the four insects tested. If NSE is to be further developed as a tree treatment, more in-depth studies will be required in which whole trees are treated. Alternative methods of application such as soil and trunk injections will need to be examined. Results with trunk injections suggest that this may be an efficacious method of applying NSE to whole trees.[19] Further work with timing, concentration, and number of NSE applications to trees is required to determine if the methods and cost of NSE application compare favorably with those of commercially available insecticides.

III. SHRUBS AND FLOWER CROP PESTS

A. JAPANESE BEETLE (*POPILLIA JAPONICA* NEWMAN)

One of the most troublesome pests of shrubs and flowers outdoors in the eastern U.S. is the Japanese beetle, *P. japonica* Newman (Coleoptera:Scarabaeidae). This insect is a polyphagous feeder of foliage and flowers as an adult and of roots as a larva.

In two separate studies, the repellency and toxicity of NSE to Japanese beetles has been studied. When sprayed or painted with NSE, both sassafras (*Sassafras albidum* [Nutt.]) and soybean (*Glycine max* [L.]) foliage was almost completely repellent to adult beetles.[20,21] When given a choice, beetles consumed all of the untreated half of a leaf but left "practically untouched" the half treated with NSE. When given no choice, beetles fed only very lightly on sassafras leaves treated with 0.5 to 2.0% NSE, and some starved rather than feed. The residual repellency of NSE-treated soybean leaves lasted 12 d, and only "moderate feeding" was oserved 17 d after spraying compared with consumption of all untreated foliage. Interestingly, repellency was not demonstrated when rose blossoms and grape foliage were sprayed; the attraction of these plants may have overridden the repellency of NSE.[22]

Toxicity of azadirachtin (AZ), one of the insecticidal constituents in NSE,[8] against the larval, prepupal, and early pupal stage of the Japanese beetle was demonstrated in the laboratory.[23,24] Nearly all prepupae died when topically treated with a range (1 to 4 μg) of AZ dissolved in ethanol. When applied to pupae 24 h after molt, AZ caused 79% mortality, but pupae that were 72 h old or older were not affected by 0.25 to 4 μg AZ. Larvae were deemed the most susceptible stage based on 90 to 100% mortality after application of 2 μg/larva. A lower dose of 1.6 μg/larva significantly increased the length of larval development compared with control insects. Treatment of larvae with AZ did not lead to immediate mortality, but caused death later as the insect molted to a pupa or adult.

Because larval stages of the Japanese beetle are subterranean,[25,26] efficacy of NSE or AZ against larvae will require soil applications. Demonstration of the NSE effectiveness against soil-inhabiting insects is needed, particularly in view of the fact that its activity is soil-type dependent.[27,28]

B. BAGWORM (*THYRIDOPTERYX EPHEMERAEFORMIS* [HAWORTH])

When Chinese junipers (*Juniperus chinensis* L.) infested with bagworms (*T. ephemeraeformis* [Haworth]) (Lepidoptera:Psychidae) were sprayed with 0.2% NSE, no larval mortality was observed 4 d after spraying.[29] Synthetic pesticides gave complete control. Long term effects of NSE over the duration of larval development were not followed.

C. BEET ARMYWORM (*SPODOPTERA EXIGUA* [HUBNER])

Larvae of beet armyworm, *S. exigua* (Hubner) (Lepidoptera:Noctuidae), feed on the foliage of a wide variety of flower and vegetable crops.[25] When Margosan-O was sprayed at concentrations between 0.2 and 40 ppm onto chrysanthemum (*Dendranthema grandiflora* Tzvelev.) foliage infested with beet armyworm larvae, there was no effect on the insect 1 d or 2 weeks after treatment.[30] Similarly, when larvae were placed on leaves that had been sprayed and allowed to dry, the residual Margosan-O had no effect on larval development. Thus although NSE has been shown to be active against other lepidopteran larvae[8] including the related *S. littoralis* (Boisd.),[31,32] a vegetable crop pest, Margosan-O was not active against the beet armyworm.

D. GREENHOUSE WHITEFLY (*TRIALEURODES VAPORARIORUM* WESTWOOD)

Greenhouse whiteflies (*T. vaporariorum* Westwood) (Homoptera:Aleyrodidae) are ubiquitous pests of most greenhouse flower crops.[33] They damage plants by feeding on vascular fluids. When chrysanthemums were sprayed with 0.4% NSE and then placed in a colony of greenhouse whiteflies for 24 h, the number of eggs laid on the plants was significantly reduced to one fourth or one fifth the number laid on water-sprayed plants in two trials.[34] A 0.2% NSE was more variable in its effect; in one trial it caused an insignificant 30% reduction in eggs laid, but in a second trial it caused a significant 56% reduction. Drenching the medium in which chrysanthemums were grown did not significantly reduce egg laying. Thus NSE sprayed on the plant served as an oviposition inhibitor, but as a systemic drench it failed to repel adults.

Simular reductions in egg laying were observed when 0.2 and 2.0% NSE were sprayed on cotton foliage against another whitefly, *Bemisia tabaci* (Gennadius).[35] Additionally, NSE reduced egg viability, prolonged larval development, and increased larval mortality in *B. tabaci,* a pest primarily of vegetable crops.

E. *LIRIOMYZA* LEAFMINER
1. Neem Seed Extract
Liriomyza leafminers, particularly *L. trifolii* (Burgess) (Diptera:Agromyzidae), are international pests of several floricultural crops including chrysanthemum, gerbera, gypso-

phila, cineraria, marigold, zinnia, and dahlia. Damage is aesthetic and is caused by larvae as they mine through foliage and, occasionally, flower petals. *L. trifolii* is difficult and expensive to control.[5]

Several laboratories have studied the use of NSE against *L. trifolii*. The first report was published in 1983,[36] and included an evaluation of the NSE antiovipositional and larvacidal activity against *L. trifolii* and *L. sativae* Blanchard, a leafminer pest on vegetable crops. When bean leaves were dipped in 0.1% NSE, significantly fewer *L. trifolii* eggs were laid in the leaves compared to water-dipped leaves. This difference, however, was not observed with *L. sativae*, and the antiovipositional activity against *L. trifolii* decreased to no activity 22 h after treatment. These variable results between species and over time indicate that NSE may not be a useful antiovipositional material for *Liriomyza* spp. The same study, however, demonstrated that both species were sensitive to the larvicidal properties of NSE. *L. sativae* experienced 95% mortality if leaves were treated with 1% NSE when leafminers were eggs or at first, second, or third instar. At a lower concentration (0.1%), NSE caused greater mortality of the larval stage than the egg and significant residual larvicidal activity was seen 7 d after treatment. Results indicated that NSE toxins move from the leaf surface to inside the leaf and disrupt development of internally feeding plant pests.

In a later study,[37] steam distillates of fresh neem leaves were used at 0.3 to 1.2% in water. Third instars extracted from bean leaves and placed on filter paper soaked in neem leaf distillates generally failed to pupate. If pupae were placed on distillate-soaked filter paper, they generally failed to emerge as adults, although the toxicity to pupae was slightly less than to larvae. These results indicate that neem toxins do not necessarily have to be ingested by immature *L. trifolii* to kill the insect. Dipping leaves in distillate also reduced the number of reared pupae, but was toxic to the hymenopterous parasite, *Hemiptarsenus semialbiclava* Girault. Sprays of neem leaf extract on field plots of leafminer-infested onions led to higher yields (bulb fresh weight) than in control plots.[37]

Other studies were conducted to determine the potential use of NSE and Margosan-O against *L. trifolii* in the commercial greenhouse.[27,38-49] Results from trials in research greenhouses established that a 0.4% NSE (9.2 ppm AZ) soil drench to potted chrysanthemums did not affect adult feeding or oviposition on treated plants, did not cause phytotoxicity, and caused complete mortality of pupae reared from treated plants for up to 21 d after drenching.[38-42,45] Drenches of 0.4% NSE applied to small plots in a commercial chrysanthemum greenhouse caused as much leafminer mortality as did sprays with cyromazine, an efficacious insecticide.[38] In all drench trials, NSE allowed the larvae to survive and damage the crop, i.e., there was a time lag between application and leafminer death. This delay was observed with two concentrations of NSE. Thus, when applied in a drench, speed of leafminer kill was not dose-related. Other researchers[50] found that a soil drench to beans was most active if applied 5 d before plants were exposed to *L. trifolii*, rather than 1 or 3 d, indicating that NSE drenches do not immediately make plants insecticidal.

Another trial indicated that when untreated nonfeeding prepupal third instar larvae were placed on NSE-drenched soil, 89% of the larvae failed to develop to adults.[38] Thus as observed previously,[37] contact with NSE in a pupation medium may suffice to kill leafminers.[38] NSE may also affect adult vigor. Fewer larval *L. sativae* were produced by pairs of adults reared from NSE-drenched tomatoes than from pairs reared from untreated tomatoes.[27]

Foliage sprays (0.4 to 0.5% NSE) were also effective against the leafminer, and caused mortality sooner, at the larval stage, than did drenches.[44] The speed of kill caused by NSE sprays was also shown to be dependent on concentration,[44] i.e., the higher the concentration the quicker the kill. Presumably, it should be possible to identify a range of nonphytotoxic[51] NSE concentrations that, when sprayed, would kill leafminer larvae before the foliage was made unattractive by third instar mines.

Efficacy of NSE against *L. trifolii* on other flower crops such as cineraria[47] and gerbera[48] has also been reported, and in the case of cineraria, NSE sprayed on foliage after oviposition led to fewer larval mines suggesting ovicidal activity of NSE.

2. Margosan-O

When tested as either a soil drench (0.33%; 9.9 ppm AZ) or foliar spray (1.25%; 37.5 ppm AZ), Margosan-O was similar in efficacy, residual activity and nonphytotoxicity to NSE both in research and commerical greenhouses.[43,44,46] Speed of kill was slow when 0.33% Margosan-O was used as a drench on beds in a commercial greenhouse, but was generally faster and was concentration dependent when used as a spray.[44,46] These results with sprays have been confirmed in studies by other scientists.[50] Both methods of application provided thorough kill so that very few adults were reared from Margosan-O-treated plants.

An interesting exception to the slowness of drenches was observed when individual potted plants in a research greenhouse were drenched with Margosan-O.[43] The treatment led to concentration-dependent speed of kill in chrysanthemums, i.e., when applied following oviposition, higher concentrations caused larval death and lower concentrations caused pupal death. The same drenches to marigolds, however, did not affect speed of kill.[49] Only pupae were killed. Thus, drench efficacy may be host plant dependent.

Margosan-O drenches did not cause significant leafminer death on zinnia or snapdragon.[43] The fact that snapdragons were poorly attacked by leafminer adults probably explains why efficacy was not detected on this crop. Results from the zinnias, however, further highlight the point that efficacy of soil-applied Margosan-O may depend on the type of plant tested. Application method may also influence efficacy. For example, in spite of poor drench efficacy on zinnias, when sprayed on zinnias, 0.5% Margosan-O caused significant pupal death.[52]

3. Other Findings

No ovicidal activity of NSE[53,54] or Margosan-O[55] was observed on chrysanthemums, but when sprayed, Margosan-O was found "to be somewhat repellent to adult feeding."[56] Other researchers found that NSE repelled adults from feeding for only 1 d.[54]

Rooted chrysanthemum cuttings placed in 0.5% NSE or 1% Margosan-O for 24 h, then grown for up to 4 weeks before being exposed to *L. trifolii* yielded significantly fewer reared adults than did water-dipped plants.[50,57] Both NSE and Margosan-O, however, stunted the growth of the cuttings. If phytotoxicity could be minimized, both materials might be useful in controlling the leafminer during plant propagation.

A survey of toxicity of neem seed extracts that were made using a variety of solvents showed that methanolic extracts were most consistently toxic to *L. trifolii* when sprayed or painted on bean plants, as opposed to extracts made with acetone, chloroform, dioxane, and other organic solvents.[58]

The movement of neem toxin in chrysanthemums and beans was studied using *L. trifolii* as the test insect.[59] Untreated chrysanthemum foliage on a plant above or below foliage sprayed with 1% NSE was not toxic to leafminers. Thus, NSE toxin did not move up or down the chrysanthemum stem in lethal concentrations away from the region of spray application. Untreated lima bean (*Phaseolus limensis* cv Henderson bush) leaves opposite to leaves painted with 0.4% NSE were somewhat toxic to leafminers, but not as toxic as leaves painted with NSE. This finding indicates that when applied topically to bean foliage, NSE toxin may move from leaf to leaf, but is diluted in transit. Foliage that was present on chrysanthemum cuttings when the cuttings were root-dipped in 0.4% NSE was toxic to leafminers, but new foliage that developed following treatment was not. Soaking bean seeds in 4% NSE for 6 h did not lead to toxic seedlings, but drenching the germination medium with 0.8% NSE did. Together, these results indicate that movement of NSE toxin when topically applied, is limited if it occurs at all, and that movement may be influenced by plant species. Complete spray coverage is required for maximum efficacy. Treatment of cuttings and seeds with NSE has limitations; drenches to potting or germination media are more effective in making cuttings or seedlings toxic to leafminers.

F. SUMMARY

Published reports of the Margosan-O activity against ornamental pests are only available for the beet armyworm and *L. trifolii*. Although results from the armyworm are not encouraging, all leafminer research indicates that the commercial product is as useful as NSE.

Research on shrub and flower pests is admittedly limited, but it has contributed an important basis of findings from which to advance. For example, the option of using NSE as either a drench or spray is available, and based on insect and plant species, method of growing, seriousness of infestation, and availability of NSE a grower would choose the appropriate application method. Generally, spraying causes quicker leafminer death at about the same concentration as comparable drenches. However, sprays, unless applied thoroughly, may fail to reach all larvae. A drench, although slower, imparts toxicity to all foliage present. With few exceptions results from shrub and flower pests are promising enough to warrant additional research.

IV. DISCUSSION

Studies of NSE used on ornamental plants are important in two respects. First, they have highlighted the prospects of using NSE in developed countries. None would argue the importance of taking full advantage of the NSE potential in underdeveloped and developing countries; as an effective alternative to expensive imported commercial insecticides, NSE holds promise in these locations. At the same time, however, we should actively pursue its use in developed countries where the need for environmentally safe control methods is pressing. The efforts of scientists and industry in developing realistic protocols for use of NSE against Japanese beetle and *L. trifolii* in commercial settings is noteworthy. They constitute some of the first successful moves from laboratory to site of use. They have begun to bring NSE down to earth in the minds of other researchers, administrators, and funding agents in the U.S. The realities of using NSE commercially in the U.S. are clearer for the efforts.

Second, the studies have served to support or amend generalizations made about neem. For example, the observations by Gill and Lewis[28] of the NSE systemic action has been pursued actively using *L. trifolii*. Likewise, NSE studies of residual activity, dose dependency, phytotoxicity, plant-mediated toxicity, and methods of application have all been advanced by work with ornamental pests.

The focus of reports of NSE used against pests of ornamentals primarily has been on practical usage with an eye to the most effective and efficient use of NSE in control programs. In that respect ornamentals research provides the groundwork, at least in developed countries, for commercial use of NSE. Presumably many of the findings derived from research on ornamentals will cue scientists working with food and fiber pests.

ACKNOWLEDGMENTS

I thank Janet Knodel, Ralph Webb, David Warthen, Martin Jacobson, and Dan Marion for information used in preparing this chapter. Thomas Elden and T. L. Ladd reviewed the manuscript.

REFERENCES

1. **Robertson, J. L.,** International Movement of Fresh Cut Flowers, Ohio State University Report, May-June 1983.
2. **Anon.,** Horticultural products, Foreign Agricultural Circular FHORT-8-85, USDA, *Foreign Agricultural Service, Washington, D.C.,* 1985.
3. **Anon.,** Flowers unlimited: international developments in floriculture, Verenigde Bloemenveilingen Aalsmeer, Aalsmeer, The Netherlands, 1982.
4. **Anon.,** Floriculture Crops: 1985 summary, intentions for 1986, Report SpCr 6-1(86), Crops Reporting Board, Stat. Rep. Serv., U.S. Department of Agriculture, 1986.
5. **Prevatt, J. W. and Price, J. F.,** An economic analysis of costs for leafminer control on two pompon chrysanthemum farms in Florida, in *Proc. 3rd Annu. Industry Conf. on the Leafminer,* Poe, S. L., Ed., 1982, 124.
6. **Lindquist, R. K.,** Special efforts to reduce survival of the resistant, *Greenhouse Manager,* December, 1985, 96.
7. **Parrella, M. P., Barker, J. R., Sanderson, J. R., Lindquist, R. K., and Ali, A. D.,** Resistance to insecticides: what a grower can do, *Grower Talks,* March 1987, 136.
8. **Warthen, J. D., Jr.,** *Azadirachta indica:* a source of insect feeding inhibitors and growth regulators, U.S. Department of Agriculture, Agric. Rev. Manuals, ARM-NE4, 1979.
9. **Ahmed, S., Grainge, M., Hylin, J. W., Mitchell, W. C., and Litsinger, J. A.,** Some promising plant species for use as pest control agents under traditional farming systems, in *Natural Pesticides from the Neem Tree and Other Tropical Plants,* Schmutterer, H. and Ascher, K. R. S., Eds., GTZ Press, Eschborn, West Germany, 1984, 565.
10. **Houston, D. R.,** Effects of defoliation on trees and stands, in *The Gypsy Moth: Research Toward Integrated Pest Management.* Doane, C. C. and McManus, M. L., Eds., Tech. Bull. 1584, Forest Service, U.S. Department of Agriculture, 1984, 215.
11. **Skatulla, U. and Meisner, J.,** Research with neem seed extract on the control of the gypsy moth, *Lymantria dispar* (L.), *Anz. Schaedlingskd., Pflanz. Umweltschutz,* 48, 38, 1975.
12. **Meisner, J. and Ascher, K. R. S.,** Extracts of neem (*Azadirachta indica*) and their effectiveness as pesticides for different insects, *Phytoparasitica,* 14, 171, 1986.
13. **Speckbacher, U.,** The effect of components of *Azadirachta indica* A. Juss on Several Species of Insects, Ph.D. dissertation, University of Munich, West Germany, 1977, 77.
14. **Feuerhake, K. J.,** Effectiveness and selectivity of technical solvents for the extraction of neem seed components with insecticidal activity, in *Natural Pesticides from the Neem Tree and Other Tropical Plants,* Schmutterer, H. and Ascher, K. R. S., Eds., GTZ Press, Eschborn, West Germany, 1984, 103.
15. **Schmutterer, H.,** Which insect pests can be controlled by application of neem seed kernel extracts under field conditions?, *Z. Angew. Entomol.,* 100, 468, 1985.
16. **Larew, H. G., Knodel, J. J., and Marion, D. F.,** Use of foliar-applied neem (*Azadirachta indica* A. Juss.) seed extract for the control of the birch leafminer, *Fenusa pusilla* (Lepeletier) (Hymenoptera:Tenthredinidae), *J. Environ. Hortic.,* 5, 17, 1987.
17. **Larew, H. G. and Knodel, J. J.,** Effect of neem (*Azadirachta indica* A. Juss.) seed extract on birch leafminer, New York, 1986, *Insecticide and Acaricide Tests,* 12, 349, 1987.
18. **Schultz, P. B. and Coffelt, M. A.,** Orange striped oakworm control on willow oak, 1985, *Insecticide and Acaricide Tests,* 11, 415, 1987.
19. **Larew, H. G., Marion, D. F., and Knodel, J. J.,** Use of pressure-injected neem (*Azadirachta indica* A. Juss.) seed extract for control of the birch leafminer, *Fenusa pusilla* (Lepeletier) (Hymenoptera:Tenthredinidae), in preparation.
20. **Ladd, T. L., Jacobson, M., and Buriff, C. R.,** Japanese beetle: extracts from neem tree seeds as feeding deterrents, *J. Econ. Entomol.,* 71, 810, 1978.
21. **Anon.,** Neem tree: seed extracts repel Japanese beetles, *Agric. Res.,* March 1979, 8.
22. **Ladd, T. L.,** Neem seed extracts as feeding deterrents for the Japanese beetle, *Popillia japonica,* in *Natural Pesticides from the Neem Tree (Azadirachta indica A. Juss.),* Schmut-

terer, H., Ascher, K. R. S., and Rembold, H., Eds., GTZ Press, Eschborn, West Germany, 1981, 149.

23. **Ladd, T. L., Warthen, J. D., Jr., and Klein, M. G.,** Japanese beetle (Coleoptera: Scarabaeidae): the effects of azadirachtin on the growth and development of the immature forms, *J. Econ. Entomol.,* 77, 903, 1984.

24. **Ladd, T. L.,** The influence of azadirachtin on the growth and development of immature forms of the Japanese beetle, *Popillia japonica,* in *Natural Pesticides from the Neem Tree and Other Tropical Plants,* Schmutterer, H. and Ascher, K. R. S., Eds., GTZ Press, Eschborn, West Germany, 1984, 425.

25. **Westcott, C.,** *Gardener's Bug Book,* Doubleday, Garden City, N.Y., 1973.

26. **Klein, M. G.,** personal communication, July 1, 1987.

27. **Webb, R. E., Larew, H. G., Weiber, A. W., Ford, P. W., and Warthen, J. D., Jr.,** Systemic activity of neem seed extract and purified azadirachtin against *Liriomyza* leafminers, in *Proc. 4th Annu. Ind. Conf. on Liriomyza Leafminers,* Poe, S. L., Ed., Sarasota, Fla., 1984, 118.

28. **Gill, J. S. and Lewis, C. T.,** Systemic action of an insect feeding deterrent, *Nature (London),* 232, 402, 1971.

29. **Schultz, P. B. and Coffelt, M. A.,** Bagworm control on juniper, 1985, *Insecticide and Acaricide Tests,* 11, 411, 1986.

30. **Harris, M., Warkentin, D., and Begley, J.,** Margosan-O contact and residual efficacy against beet armyworm larvae, 1986, *Insecticide and Acaricide Tests,* 12, 331, 1987.

31. **Meisner, J., Ascher, K. R. S., Aly, R., and Warthen, J. D., Jr.,** Response of *Spodoptera littoralis* (Boisd.) and *Earias insulana* (Boisd.) larvae to azadirachtin and salannin, *Phytoparasitica,* 9, 27, 1981.

32. **Meisner, J., Ascher, K. R. S., and Zur, M.,** The residual effect of a neem seed kernel extract sprayed on fodder beet against larvae of *Spodoptera littoralis, Phytoparasitica,* 11, 51, 1983.

33. **Vittum, P.,** Greenhouse whitefly, University of Massachusetts Coop. Ext. Serv. Entomol. Bull. No. 6, 1982.

34. **Larew, H. G. and Knodel, J. J.,** Effect of neem *(Azadirachta indica* A. Juss.) seed extract on whitefly oviposition, Insecticide and Acaricide Tests,11, 377, 1986.

35. **Coudriet, D. L., Prabhaker, N., and Meyerdirk, D. E.,** Sweet potato whitefly (Homoptera:Aleyrodidae): effects of neem seed extract on oviposition and immature stages, *Environ. Entomol.,* 14, 776, 1985.

36. **Webb, R. E., Hinebaugh, M. A., Lindquist, R. K., and Jacobson, M.,** Evaluation of aqueous solution of neem seed extract against *Liriomyza sativae* and *L. trifolii* (Diptera:Agromyzidae), *J. Econ. Entomol.,* 76, 357, 1983.

37. **Fagoonee, I. and Toory, V.,** Contribution to the study of the biology and ecology of the leaf-miner *Liriomyza trifolii* and its control by neem, *Insect Sci. Appl.,* 5, 23, 1984.

38. **Larew, H. G., Knodel-Montz, J. J., Webb, R. E., and Warthen, J. D.,** *Liriomyza trifolii* (Burgess) (Diptera:Agromyzidae) control on chrysanthemum by neem seed extract applied to soil, *J. Econ. Entomol.,* 78, 80, 1985.

39. **Larew, H. and Knodel-Montz, J.,** The efficacy of neem seed products for the control of *Liriomyza trifolii* (Burgess) (Diptera:Agromyzidae), in *Proc. 1st Conf. on Insect and Mite Manag. Ornamentals,* Parrella, M. P., Ed., 1985, 34.

40. **Larew, H. G., Webb, R. E., and Warthen, J. D., Jr.,** Leafminer controlled on chrysanthemum by neem seed extract applied to potting soil, in *Proc. 4th Annu. Industrial Conf. on Leafminer,* Poe, S. L., Ed., Sarasota, Fla., 1984, 108.

41. **Larew, H. G., Knodel-Montz, J. J., and Webb, R. E.,** Neem seed extract products control a serpentine leafminer in a commercial greenhouse, in *Proc. Informal Conf. on Liriomyza Leafminers,* Knodel-Montz, J. J., Ed., 1985, 33.

42. **Larew, H. G. and Knodel-Montz, J. J.,** Control of leafminer using neem *(Azadirachta indica* A. Juss.) drenches, *Insecticide and Acaricide Tests,* 10, 282, 1984.

43. **Knodel-Montz, J. J., Larew, H. G., and Webb, R. E.,** Efficacy of Margosan-O, a formulation of neem, against *Liriomyza trifolii* (Burgess) on floral crops, in *Proc. Informal Conf. on Liriomyza Leafminers,* Knodel-Montz, J. J., Ed., 1985, 33.

44. **Knodel-Montz, J. J. and Larew, H. G.,** Control of leafminer using neem *(Azadirachta indica* A. Juss.) products and other insecticides as foliar sprays, *Insecticide and Acaricide Tests,* 10, 282, 1985.

45. **Larew, H. G. and Knodel, J. J.,** Test for plant growth responses using neem *(Azadirachta indica* A. Juss.) as a soil drench, *Insecticide and Acaricide Tests,* 10, 306, 1984.

46. **Knodel, J. J., Larew, H. G., and Webb, R. E.,** Margosan-O™, a commercial formulation of neem seed extract, controls *Liriomyza trifolii* on chrysanthemums, *J. Agric. Entomol.,* 3, 249, 1986.

47. **Larew, H. G. and Knodel, J. J.,** Control of leafminers (on *Cineraria*) using neem *(Azadirachta indica* A. Juss.) seed extract, 1985, *Insecticide and Acaricide Tests,* 11, 383, 1986.

48. **Larew, H. G. and Knodel, J. J.,** Control of

leafminers (on *Gerbera*) using neem (*Azadirachta indica* A. Juss.) seed extract, 1985, *Insecticide and Acaricide Tests*, 11, 389, 1986.

49. **Knodel, J. J. and Larew, H. G.,** Leafminer control using Margosan-O, a neem formulation (*Azadirachta indica* A. Juss.) seed product on marigolds, 1985, *Insecticide and Acaricide Tests*, 11, 414, 1986.

50. **Lindquist, R. K., Krueger, H. R., and Casey, M. L.,** Use of neem extract as a systemic treatment for *Liriomyza trifolii* control on greenhouse chrysanthemum, *British Crop Prot. Conf.—Pests and Disease,* Brighton, England, 1986, 271.

51. **Knodel, J. J. and Larew, H. G.,** Test for plant growth responses using neem (*Azadirachta indica* A. Juss) products as foliar sprays, 1984, *Insecticide and Acaricide Tests,* 11, 420, 1986.

52. **Knodel, J. J. and Larew, H. G.,** Leafminer control using Margosan-O, a neem (*Azadirachta indica* A. Juss) seed product on zinnias, 1985, *Insecticide and Acaricide Tests,* 11, 432, 1986.

53. **Stein, U.,** The Potential of Neem for Inclusion in a Pest Management Program for Control of *Liriomyza trifolii* on Chrysanthemum, M.S. thesis, University of California, Riverside, 1984.

54. **Stein, U. and Parrella, M. P.,** Seed extract shows promise in leafminer control, *Calif. Agric.,* July-August, 19, 1985.

55. **Harris, M., Warkentin, D., and Begley, J.,** Margosan-O ovicidal and larvicidal efficacy against leafminer, 1986, *Insecticide and Acaricide Tests,* 12, 330, 1987.

56. **Warkentin, D. and Harris, M.,** Careful planning needed to avoid leafminers, *Grower Talks,* 51, 52, 1987.

57. **Lindquist, R. K.,** Natural pesticides could be irresistible, *Greenhouse Manager,* November, 110, 1986.

58. **Meisner, J., Yathom, S., Tal, S., and Ascher, K. R. S.,** The effect of various extracts of neem seed kernel on *Liriomyza trifolii* (Burgess) (Diptera:Agromyzidae), *J. Plant Dis. Prot.,* 93, 146, 1986.

59. **Larew, H. G.,** Limited occurrence of foliar-, root-, and seed-applied neem seed extract toxin in untreated plant parts, *J. Econ. Entomol.,* 81, 593, 1987.

6

Effects of Neem on Stored Grain Insects

R. C. Saxena, G. Jilani, and A. Abdul Kareem
Department of Entomology
The International Rice Research Institute
Manila, Philippines

TABLE OF CONTENTS

I. INTRODUCTION

It is generally accepted that settled agriculture began about 10,000 years ago and the practice of storing food grains started about 4500 years ago as a safeguard against poor harvests and famines. There is evidence that several species of storage insect pests occurred in granaries and other food storage structures in ancient times.[1] Insect pests cause heavy foodgrain losses in storage, particularly at the farm level in tropical countries. The increased foodgrain production due to the introduction of high yielding varieties in the last 2 decades has further accentuated the storage problem. Despite the current glut in food production, buffer stocks of foodgrains are needed in most of the developing countries.

Although foodgrains are now commonly protected by insecticidal application or fumigation, such practices are health risks unless the chemicals used are inherently safe to man. Toxic residues and the selection of insecticide-resistant pest strains are additional problems associated with the use of insecticides in grain protection. At least 447 insect and mite species, including stored grain pests, have become resistant to one or more classes of insecticides.[2] Nonetheless, the continued need for maintaining buffer stocks will not permit significant reduction of pesticide use. However, the new pesticides will have to meet entirely different standards. They must be pest-specific, nontoxic to man, biodegradable, less prone to pest resistance, and relatively less expensive. This has led to re-examining the century-old practices of protecting stored products using derivatives of some plants which have been known to resist insect attack. It is now recognized that insects fail to damage these plants because they possess a complex array of inherent defense chemicals. Thus many of these plant species are potential sources of botanical pesticides. In fact, many of the oldest and most common pesticides, such as nicotine, pyrethrins, and rotenone, were derived from plants. The chemical or pesticide approach had its beginning in the use of botanical materials.[3]

In recent years, neem *Azadirachta indica* A. Juss, has been widely investigated as a source of natural insecticides.[4-8] This tropical tree is widespread in Asia and Africa and has long been known to be free of insects, nematodes, and plant diseases; it thrives even under adverse climatic conditions. The scientific name of the tree is derived from "azad dirakht-i-hind" which in Persian, Urdu, and Hindustani languages means "free tree of India".[9] The bitter taste in almost every part of the neem tree discourages insect attack and renders it pest-free.

II. TRADITIONAL USE OF NEEM AS A GRAIN PROTECTANT

It has been an age-old practice in rural India to mix dried neem leaves with stored grains and to place leaves in the folds of woolen clothes to ward off insect attack. This practice has been inherited as a tradition and as part of the culture. Pruthi and Singh[10] recorded that neem leaves were spread in 5- to 7-in. thick layers in grains and neem fruits were crushed on the inner surfaces of grain cribs. Mixing of neem leaves with wheat, rice, or other grains (2 to 5%) is even now a common practice in many villages in India and Pakistan. Other common practices include mixing neem leaf paste with the mud that is used for making earthen containers for storing grains and overnight soaking of empty gunny bags in a 2 to 10% neem leaf decoction, using them after drying for storing grain.

The traditional use of neem may differ in different regions or with farmers of different cultural backgrounds. For example, in southern Sind, Pakistan, farmers mix neem leaves with grains stored in gunny bags, or they rub crushed neem leaves on the inner surfaces

FIGURE 1. "Palli", a grain storage structure made of plant-material and plastered with mud, is commonly used by farmers in Sind, Pakistan.

of mud bins before filling them with grains. In central Sind, where "palli" made of plant materials is a common storage structure (Figure 1), crushed neem leaves are mixed with mud and used as plaster for its inner sidewalls and top. In southern Punjab, neem leaf extract is sprinkled on wheat straw packed at the bottom of "palli" 2 to 3 d prior to storing grain. A survey of various types of on-farm storage practices revealed that a combination of two or three insect control measures, including the use of neem leaves, was used by 29% of the farmers in Punjab and 47% of the farmers in Sind.[11] In Sri Lanka, farmers burn neem leaves to generate smoke for fumigation against insect pests that attack stored paddy and pulses.[12] In Nigeria, the traditional use of plant derivatives, including neem, for protecting stored grain has also been documented.[13-15]

The traditional use of neem derivatives is simply based on experience and the understanding that comparatively less damage occurs in the treated stored commodity. However, on scientific ground, it is clear that in traditional uses, very large quantities of neem materials are involved which presumably affect the storage environment. They produce a repellent odor which makes the insect restless and their bitter taste renders treated grain unpalatable to insects. However, the exact quantity which gives the optimum effect is hardly ever known. It is therefore desirable to know the amount of various materials that provide adequate protection. It is also important to know how neem derivatives affect insect behavior, growth and development, and reproduction. Further improvements can be brought about by using enriched neem extracts or active fractions and finding better methods of application to stored products.

The active principals or the "bitters" in neem and the related Persian lilac (chinaberry), *Melia azedarach* L., have been identified as limonoids, a group of stereochemically homogeneous tetranorterpenoids.[16-24] These limonoids occur predominantly in the family Meliaceae and also in Rutaceae. Azadirachtin, the best known neem and chinaberry constituent, has shown selective activity against many insect pest species affecting agriculture and health. In general, insects feed less, grow poorly, and lay fewer eggs on treated substrates. Toxicity of neem derivatives against storage insects has also been reported. However, insect mortality in such cases was attributed to starvation, exhaustion, or growth disruption, rather than direct contact toxicity. This chapter reviews primarily the repellent, antifeedant, oviposition deterrent, and growth and reproduction inhibitory effects of neem on stored product insects. On-farm studies against storage insects are also briefly reviewed.

FIGURE 2.   Food preference chamber used for evaluating repellency of treated grain.[37] The platform has 12 holes holding alternating treated or untreated grain samples in petri dishes.

III. REPELLENT EFFECTS

An insect repellent is a chemical stimulus which causes the insect to make oriented movements away from the source of stimulus.[25] In many situations, it is likely that the use of repellents will affect a number of behavioral reflexes involved, sometimes sequentially, sometimes simultaneously. Immediate responses to repellents may comprise many behavioral manifestations of the interaction of chemicals with exteroceptive sense organ of the pest organism. The senses primarily involved in the detection of chemicals are olfactory or gustatory chemoreceptors.

Pruthi[26] reported the efficacy of neem leaves and neem cake as effective repellents for pests in stored wheat. Repellent action of neem leaves was also reported in trials in which neem leaves were mixed with cacao beans against the almond moth (*Cadra cautella* [Walker]),[27] the cowpea weevil (*Callosobruchus maculatus* [F.]),[28] and the rice weevil (*Sitophilus oryzae* [L.]).[29] Among several indigenous plants and plant products tested, neem derivatives were the most effective insect repellents.[30-31] Neem seed kernel powder was reported as a deterrent against the lesser grain borer (*Rhyzopertha dominica* [F.]), *S. oryzae*, and the khapra beetle (*Trogoderma granarium* Everts) in wheat and against *C. maculatus* in legumes.[32] Neem seed kernel powder mixed with paddy at 2% effectively repelled *R. dominica* and at 3% repelled *S. oryzae*.[33] Mixing neem products did not affect the cooking quality of grain and germination. Neem seed powder mixed with red corn at 2, 4, or 8% repelled the confused flour beetle (*Tribolium confusum* Jacquelin du Val) and the maize weevil (*S. zeamais* Mots.).[34] Smoke generated by burning neem leaves has also been reported to banish grain moths from warehouses.[35]

Many of these studies lacked standardized techniques for testing repellency. Recently, Jilani and Su[36] have demonstrated the repellency of neem leaf powder against the granary weevil (*S. granarius* [L.]) and *R. dominica* using standardized techniques. Neem leaf powder was mixed with undamaged soft winter wheat (1 or 2% by weight) and evaluated by the repellency wheel method by using the modified Loschiavo food preference apparatus described by Laudani and Swank.[37]

The apparatus comprises a 50-cm-diameter circular platform with a 5-cm metal rim (Figure 2). The platform has 12 holes (8.75-cm diameter) equally spaced along the outer edge to accommodate paper cups or plastic petri dishes which are filled to the brim with alternating treated or untreated grain sample. The center of the wheel cover

FIGURE 3. Paper strip method for testing repellency.[43] The test arena inside the glass rings comprises treated and untreated halves exposed to the test insects.

has an opening (1.25-cm diameter) fitted with a tube through which test insects are introduced for a 24-h exposure period. At the end of the exposure period, the insects in the cups or the petri dishes are counted. Likewise, repellency of neem seed kernel powder[38] or of neem oil[39] mixed with milled rice against the red flour beetle (*T. castaneum* [Herbst]) has been demonstrated.

Neem seed extracts are even stronger repellents against storage insects. Petroleum ether extract mixed with wheat strongly repelled *S. oryzae* and *T. castaneum*.[40] Neem seed extract mixed at 1 to 3% with bengal gram effectively repelled *C. maculatus*.[41] Neem oil mixed with red corn at 1 to 8 ml/kg repelled *T. confusum* and *S. zeamais*.[34] Repellency of aqueous and alcoholic extracts of neem leaves, flowers, unripe fruit, and seed was demonstrated against *T. castaneum* larvae and *R. dominica* and *Trogoderma granarium* adults by alternately using treated and untreated filter paper strips in a test arena.[42] Neem seed extract was the most repellent one. The petroleum ether extract of neem leaves was a stronger repellent against *Tribolium castaneum*[36] than acetone or ethanol extract.

The repellency test was conducted using the method described by Laudani et al.[43] and McDonald et al.[44] Strips of aluminum foil laminated to 40-lb kraft paper are treated on the paper side with acetone solution of the extract at desired concentrations and then air dried. Each treated strip (10 × 20 cm) is then attached lengthwise, edge to edge, to an untreated strip by cellulose tape on the reverse side. Two glass rings, 2.5 cm high and 6.4-cm diameter, are placed over the two matched papers so that the joined edges bisect the rings (Figure 3). The insects are exposed on each test arena inside the glass ring and their numbers on the treated and untreated halves are recorded at definite time intervals for 5 consecutive days. Average insect count of eact 3-d period is converted to percent repellency by deducting percent individuals on treated half from those in the control. Similarly, repellency of azadirachtin against *T. castaneum*[45] and of neem oil and "Margosan O" (an azadirachtin-rich commercial formulation) against *T. castaneum* and *R. dominica* has been demonstrated.[39]

IV. ANTIFEEDANT EFFECTS

The chemicals which retard or disrupt insect feeding by rendering the treated materials unattactive or unpalatable are known as antifeedants.[46] They are not necessarily toxic to insects but can prevent or reduce feeding when perceived. Consequently, insect growth, development, survival, and reproduction are adversely affected.[47]

There are several reports on the antifeedant effects of neem derivatives on stored grain insect pests. Neem seed powder mixed with wheat grain at 1 or 2% protected it from feeding damage by *S. oryzae, R. dominica*, and *Trogoderma granarium* for about 9, 11, and 13 months, respectively.[48] The same treatment also protected mungbean, bengal gram, cowpeas, and peas against *C. maculatus* feeding for about 8 to 11 months.[49] Likewise, mixing of neem seed kernel powder at 2% protected grain sorghum from feeding damage by *S. oryzae* and at 1 to 2.5%, protected mungbean against *C. maculatus* for 6 months.[50]

Efficacy of 2% neem seed kernel powder treatments in maize grain was compared with that of ethylene dibromide-carbon tetrachloride (ED-CT) (at 30 ml/40 kg of maize) against *S. oryzae* and *Corcyra cephalonica* (Staint.).[51] The treatments were direct application to the grain in the gunny bags, application to the gunny bags, and a combination of the two. Damage at 6 months after storage was 13.1% in the grain treated with neem powder, 13.3% in the grain and bags treated with neem powder, 20.9% in those with ED-CT, 25.1% in bags treated with neem powder, and 33% in the untreated control. The corresponding weight loss in the grain was 6.1, 7.4, 10.1, 11.5, and 13%.

Neem seed powder mixed with wheat at 0.5, 1, or 2% protected wheat grain against *T. granarium* larvae.[52] The efficacy was on par with 1 to 2% magnesium carbonate treatment up to 4 to 5 months. Thereafter, only the neem treatment was effective up to 8 months. In paddy, treatment with 2% neem kernel powder during 6 months of storage was as effective as treatment with 0.01% malathion dust.[53] Neem kernel powder mixed with pigeon pea and mungbean at 2% protected them against *Callosobruchus maculatus*.[54]

In a series of trials, the efficacy of neem kernel powder was tested.[55] A treatment of 2% neem kernel powder mixed with cowpea protected it from *Callosobruchus* sp. infestation and damage for 9 months. The same treatment fully protected hybrid sorghum seed against *S. oryzae* for 3 months. After a 4-month storage period, neem-treated sorghum had 21.8% damage as against 100% in the untreated control. In bengal gram and red gram, treatment with 3% neem kernel powder kept them free of *Callosobruchus* spp. infestation for 10 months; wholesale damage occurred in the untreated control. Neem-treated sorghum seed was free from insect infestation for 8 months. There were 50 and 100% damages in the control in 5- and 6-month storage, respectively.

Wheat treated with 1, 2, or 4% powdered neem seed had substantially less *T. granarium* damage for 7 to 16 months than in the untreated control.[56] The effect increased with increasing quantities of neem powder applied. Neem seed kernel powder mixed at 2% with pigeon pea or green gram protected them from *C. chinensis* (L.) damage for about 3 months.[57] Neem seed powder at 1 to 2% was also reported to reduce bruchids and *T. granarium* damage in pulses and cereals for a considerable period.[58] Likewise, mixing of 1 to 2% neem kernel powder in cowpea protected it from *C. maculatus* damage for 8 months.[59] Neem oil at 0.8% protected cowpea seed up to 3 months; at 0.5%, it protected bambara groundnut up to 6 months.[60] A variety of neem extracts at 0.5 and 1% or neem oil at 1% protected stored paddy against *R. dominica* for 6 months.[61] Neem extract at 1% was also effective against *Sitotroga cerealella* (Oliv.).

To demonstrate the antifeedant effect of neem against *Tribolium castaneum*, Jilani and Malik[42] used P^{32}-labeled wheat flour which had been treated with neem seed extract. Radioactive-labeled wheat flour without neem treatment served as the control. The amount of radioactivity measured in the insect that fed on the flour indicated the relative food intake. The difference between the radioactivity counts in insects that fed on untreated and neem-treated flour represented the antifeedant effect due to neem. For instance, neem extract at 0.16% reduced relative food intake by 25 times in 24 h. In 8 d, intake of neem-treated flour was 21 times less than that in the control.

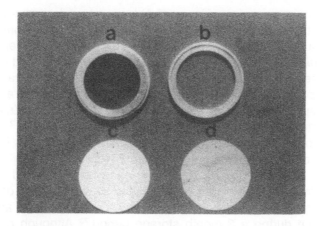

FIGURE 4.   Device for testing filter paper penetration by *Rhy-zopertha dominica*;[62] (a) = base ring and center ring with a filter paper disk in between, (b) = upper ring, (c) = treated filter paper disk, and (d) = untreated filter paper disk.

Using a simple device, resistance to penetration through neem-treated filter paper has also been used as a parameter to determine the antifeedant effect of neem against *R. dominica*.[36,62] The device comprises a base ring, a center ring, a wire mesh disk, and an upper ring (identical to the base ring) (Figure 4). After inserting a filter paper disk (7-cm diameter) into the base ring, the center ring is placed on the filter paper. The outer surface of the center ring is fitted into the flange of the base ring. After introducing ten *R. dominica* adults, the wire mesh disk is placed on the center ring and the upper ring is fitted in place in the flange to confine them. The antifeedant effect was evaluated on the basis of the number of holes made by the insect after 18, 24, 48, and 72 h. The number of feeding punctures in the filter paper disks treated with 0.5 or 1% neem extract was significantly less than that in untreated disks.

V. REPRODUCTION AND GROWTH INHIBITION

Neem derivatives have been found to adversely affect various aspects of reproduction and growth and development of stored grain pests such as grain moths, beetles, and weevils. Fry[27] reported that neem leaves mixed with cocoa beans discouraged *Cadra cautella* from laying eggs. Also, newly emerged larvae were unable to enter the sacks containing the treated beans. Mixing of neem seed or leaves also reduced the progeny production of *C. cautella* and *S. cerealella* by disturbing larval development and adult fecundity in stored maize.[63]

However, mixing of neem seed powder was not effective against *C. cautella* infestation in stored bajra (*Pennisetum typhoideum*).[64] Neem seed kernel mixed directly at 1 to 2% with paddy reduced oviposition by *S. cerealella*, while powdered neem seed kernel reduced adult emergence even at lower rates.[33,65] Neem extract or neem oil at 0.005% significantly reduced larval and adult population after the completion of one generation in treated paddy. At 0.025%, pest multiplication was completely checked.[66,67]

Application of neem oil at 0.1% to wheat grain greatly reduced egg laying by *S. cerealella* and was comparable to the treatment with 5% malathion dust.[68] Egg hatching and adult emergence in neem-treated wheat was also as low as in malathion-treated wheat. Deoiled neem kernel at 2% mixed with wheat flour arrested the growth and development of *Corcyra cephalonica* larvae.[69] At lower rates, deoiled neem kernel caused high larval mortality, reduced size and weight, delayed the development, and caused larval-pupal and pupal-adult intermediates or deformed adults. The reproductive potential of *C. cephalonica*, as expressed by the number of eggs laid and egg viability, was also significantly reduced when the pest was exposed to neem oil vapors for about 30 min.[70] Topical application of purified neem seed fractions or azadirachtin (at 0.2 to 5 μg/larva) to the spinning stage of last instar *Ephestia kuhniella* Zell. larvae caused delayed development and pupal and adult stage malformations.[71,72] At higher doses, the larvae failed to pupate even 5 weeks after treatment or died as larval-pupal intermediates. In the controls, adults emerged within 2 weeks.

Neem kernel at 1 to 2% mixed directly with paddy also reduced oviposition by *R. dominica*, while powdered neem kernel reduced the adult population of the pest at even lower rates.[33] Also, neem seed powder at 1, 2.5, or 5% reduced adult *R. dominica* population in wheat during a 3-month storage period.[73] Although neem extract was much less toxic than DDT to *R. dominica*,[74] the extract or oil applied at 0.005% to paddy significantly reduced larval and adult population after the completion of one generation.[66] At higher rates, neem extract or oil absolutely checked pest multiplication. Neem seed powder at 0.25, 0.5, or 1% mixed with wheat grain markedly reduced the emergence of *R. dominica* adults in treated wheat.[65]

Neem oil was nontoxic to *T. castaneum*,[75] but neem leaf powder at 2% mixed with wheat seed suppressed its population increase.[76] Progeny production was inhibited when *T. castaneum* adults were fed wheat flour treated with 0.04, 0.08, or 0.16% of neem seed water extract. Deoiled neem kernel powder mixed at 2% with wheat flour also arrested the growth and development of *T. castaneum* larvae.[69] The beetle failed to reproduce in wheat flour impregnated with 0.4% neem oil, in contrast to a progeny of 3509 individuals produced by 50 adults in 20 d in untreated flour.[77] The beetle fed wheat flour treated with 0.02% Margosan-O or 0.02% neem oil produced fewer and underweight larvae, pupae, and adults compared with those in the control (Figure 5).[39] Against *T. castaneum*, synergistic interaction occurred between neem oil and malathion, quinalphos, and monocrotophos insecticides but neem oil antagonized the effect of pyrethrin oil.[78]

Neem seed powder at 0.5, 1, or 2% mixed with wheat seed checked *Trogoderma granarium* multiplication and damage even upon repeated infestation with larvae up to 8 months.[52] Both nondiapausing and diapausing *T. granarium* larvae failed to develop in wheat grain treated with neem leaf or seed powder.[79] The seed powder was more effective than the leaf powder. Neem seed powder at 1, 2, or 4% mixed with wheat seed retarded the development of larvae during pupal or adult stage.[56] The effect persisted from 8 to 16 months.

Neem leaf or seed powder mixed with stored maize effectively reduced the progeny production of *Sitophilus oryzae* by disrupting larval development and adult fecundity.[63] Neem seed powder at 1% mixed with stored wheat reduced the *S. oryzae* adult progeny to almost half of that in the untreated grain.[65] Neem oil at 0.5% mixed with rice reduced *S. oryzae* and *S. zeamais* populations by almost 90%.[80] Neem seed powder at 2.5, 5, or 12.5% mixed with maize grain reduced oviposition by *S. oryzae* and stopped its postembryonic development.[81] Even after a 4-month storage period, fresh weevils placed on maize treated with 12.5% neem powder failed to oviposit and most of them died within 10 d. At 5% neem powder, progeny emergence was delayed for 6 months and the number of emerging weevils was greatly reduced. Neem treatment did not affect viability of maize seed. Powdered neem cake also reduced *S. oryzae* oviposition in treated maize grain.[82] Neem leaf or seed powder at 5% mixed with wheat grain reduced *S. oryzae* weevil population 12 to 15 times.[83] The growth inhibitory effect of neem was attributed to azadirachtin isolated from neem.[84]

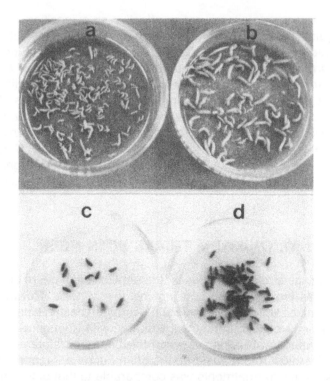

FIGURE 5. *Tribolium castaneum* larvae and adults in progeny
recovered from neem oil-treated (a, c) or untreated (b, d) wheat
flour.[97]

Neem derivatives have been widely used to protect stored pulses and other leguminous grains from pulse beetles *Callosobruchus* spp. Neem kernel powder at 1% mixed
with grass pea (*Lathyrus sativum*) provided protection against *C. chinensis* and *C.
maculatus* for up to 9 months.[85] The reduced infestation was attributed to the oviposition-
inhibitory effect of neem kernel powder. No *C. chinensis* or *C. maculatus* progeny was
produced when neem kernel powder at 2% was mixed with masur, lentil (*Lens escu-
lenta*), Bengal gram (*Cicer arietinum*), or red gram (*Cajanus cajan*). The treatment
remained effective against reinfestation for 1 year. *Callosobruchus maculatus* was more
sensitive than *C. chinensis* to neem treatment. The effect of mixing 1, 2 or 3% neem
kernel powder as seed treatment against pulse beetle in pea (*Pisum sativum*), Bengal
gram, red gram, green gram (*Vigna radiata*), black gram (*V. mungo*), lentil, and grass
pea was also evaluated.[86] In general, neem treatment inhibited progeny development
and protected leguminous grains, except pea and green gram, up to 3 to 12 months.
Newly emerged larvae were unable to feed beyond the "pin head" feeding stage on
seed coat and failed to survive. Dressing of cowpea (*V. unguiculata*) seed with 5 to
15% neem seed powder reduced the fecundity, preimaginal development, and consequently, the emergence of *C. maculatus* progeny.[87]

Neem oil at 1% applied to red gram seed as a surface protectant stopped the production of progeny by *C. chinensis* adults.[88] Even 0.1% neem oil application completely
inhibited oviposition by the beetle in treated chick pea.[89] After 1 and 3 months, infestation
in oil-treated seed was nil, while it was 26% and 99%, respectively, in the control. Oil
at 1% inhibited oviposition and caused total failure of hatchability of eggs and development of *C. chinensis* larvae in treated green gram.[90] Neem oil at 0.5% inhibited
oviposition and hatchability of eggs of *C. chinensis, C. maculatus,* and *C. analis* in green

gram;[91] at lower rates, neem oil was not as effective an oviposition inhibitor against *C. maculatus* as against the other two species. Application of neem oil at 0.8% to cowpea was more effective in inhibiting *C. maculatus* oviposition and larval development than Karite (*Butyrospermum parkii*) oil, ground nut oil, and palm oil.[60] *C. maculatus* adults laid fewer eggs and made fewer holes in cowpea seeds treated with 0.5% neem oil than in groundnut oil-treated seeds.[92] Compared with nonedible oils of Karanja (*Pongamia glabra*), undi (*Calophyllum inophyllum*), and kusum (*Schleichera trijuga*), neem oil at 0.5% was the best surface protectant against *Callosobruchus maculatus* and *C. chinensis* in green gram.[93] Fecundity and progeny development of *C. chinensis* adults reared on green gram treated with 2.5 or 5% neem leaf extract was significantly reduced and was comparable to that in green gram treated with the JH analogs hydroprene (ZR-512) or MV-678.[94]

VI. ON-FARM TRIALS WITH NEEM

At farm level storage and in warehouses, the application of neem derivatives to bags or stored grains has provided protection against insect pests. Powdered neem seed kernel at 1 to 2% mixed with paddy significantly reduced pest infestation in godowns.[53] Neem leaves at 2% mixed with paddy, bags treated with 2% neem extract, or a 20- to 30-cm-thick dried neem leaf barrier between the bags and storage floor significantly reduced insect infestation and damage to the grain during a 3-month storage period.[95] The effectiveness of neem treatments was comparable to that of 2% methacrifos dust. Likewise, neem seed extract at 7.2 g/90 kg capacity jute bag (100 × 60 cm) controlled 80% of the population of major insects and checked the damage to stored wheat up to 6 months.[96] The treatment was effective up to 13 months and provided more than 70% protection as compared to the untreated control. The neem seed extract treatment was as effective as 0.0005% pirimiphos methyl mixed with the grain. Using this technology in Sind, Pakistan, benefit-cost ratios of 4.6, 5.6, and 7.4 were attained by small-, medium-, and large-scale farmers, respesctively.[97]

VII. FUTURE CONSIDERATIONS AND CONCLUSION

Neem derivatives have been found to be effective against 123 species of insect pests, including those infesting stored products.[8] Neem derivatives are considered remarkably safe to man and warm-blooded animals. Indeed, neem cake has been fed to cattle in India in large quantities and for long periods without any deleterious effects. In a standard Ames Test, azadirachtin showed no mutagenic activity on four strains of *Salmonella typhimurium* (Loeffer) Castellani and Chalmers.[98] In subacute dermal toxicity tests, no overt signs of toxicity or abnormal behaviour were observed in albino rats administered a daily dose of "Neemrich - 100" (technical grade neem oil) at 200, 400, or 600 mg/kg.[99] The age-old traditional use of neem derivatives for protection of stored commodities demonstrates their potential as effective and safe grain protectants. Although neem derivatives degrade rapidly in the sunlight and are therefore less effective against crop pests,[100] they persist long enough in storage environments to provide adequate protection without requiring repeated applications.

The active principals in neem comprise a complex array of novel chemicals which have diverse behavioral and physiological effects on insects — repellency, feeding and oviposition deterrency, and reproduction and growth inhibition. Unlike ordinary pesticides based on a single active ingredient, the possibility of a pest developing resistance to an array of chemicals is less likely. The continued successful use of neem derivatives in several countries indicates that they continue to be effective against insect pests.

The complexity of chemical structures of neem compounds, such as azadirachtin, precludes their synthesis on a practical scale. Therefore, the use of simple formulations or derivatives such as neem leaf or kernel powder or their extracts need to be popularized. However, systematic studies are still needed to further elucidate the mode of action of neem compounds. To promote the use of neem derivatives in grain protection, ready-to-use formulations, such as dispensers, should be developed.

Collection and processing of neem seed and foliage need to be undertaken on an organized scale. Also, more neem trees will have to be propagated to ensure year-round availability of the raw material. Unlike before, there is now a widespread resurgence of interest in the pest control potential of neem. The future of neem derivatives as grain protectants looks promising as it will be supported by concerted research on their chemistry and mode of action.

ACKNOWLEDGMENTS

We gratefully acknowledge the financial grants received from the Asian Development Bank and the Swiss Development Cooperation for supporting research on the pest control potential of neem and other botanicals. We thank Ms. T. Rola for editing of the manuscript and appreciate Ms. Cecille Salonga for skillful typing of the text.

REFERENCES

1. **Levinson, H. Z. and Levinson, A. R.,** Storage and insect species of stored grains and tombs in ancient Egypt, *Z. Angew. Entomol.,* 100, 321, 1985.

2. **Georghiou, G. P.,** The magnitude of the resistance problem, in *Pesticide Resistance: Strategies and Tactics for Management,* National Academy Press, Washington, D.C., 1986, 14.

3. **Saxena, R. C.,** Naturally occurring pesticides and their potential, in *Chemistry and World Food Supplies: The New Frontiers, Chemrawn II,* Shemilt, L. W., Ed., Pergamon Press, Oxford, 1983, 143.

4. **Schmutterer, H., Ascher, K. R. S., and Rembold, H.,** Natural pesticides from the neem tree (*Azadirachta indica* A. Juss), *Proc. 1st Int. Neem Conf.,* Rottachegern, GTZ Press, Eschborn, West Germany, 1981.

5. **Schmutterer, H. and Ascher, K. R. S.,** Natural pesticides from the neem tree (*Azadirachta indica* A. Juss) and other tropical plants, *Proc. 2nd Int. Neem Conf.,* Rauischholzhauzen, GTZ Press, Eschborn, West Germany, 1984.

6. **Schmutterer, H. and Ascher, K. R. S., Eds.,** *Proc. 3rd Int. Neem Conf.,* Nairobi, Kenya, GTZ Press, Eschborn, West Germany, 1987.

7. **Jacobson, M.,** Control of stored product insects with phytochemicals, in *Proc. 3rd Int.* *Working Conf. on Stored Prod. Entomol.,* Kansas State University, Manhattan, Kansas, October 1983 (Publ. 1984), 183.

8. **Jacobson, M.,** The neem tree: natural resistance par excellence, in *Natural Resistance of Plants to Pests: Roles of Allelochemicals,* Green, M. B. and Hedin, P. A., Eds., American Chemical Society, Washington, D.C., 1986, 220.

9. **Ahmed, S.,** After Bhopal, what?, *The Economic Times,* 1, (191), Bangalore, India, April 13, 1986.

10. **Pruthi, H. S. and Singh, M.,** Stored grain pests and their control, *Imperial Council of Agric. Res.,* Misc. Bull. No. 57, 1944.

11. **Borsdorf, R., Foster, K., Huyser, W., Pedersen, J., Pfost, H., Stevens, H., and Wright, V.,** Post-harvest management project design, Pakistan, *Food and Feed Grain Inst. Rep. No. 91,* Manhattan, Kansas, 1983.

12. **Ranasinghe, M. A. S. K.,** Neem and other promising botanical pest control materials from Sri Lanka, in *Proc. Int. Workshop on the Use of Botanical Pesticides,* International Rice Research Institute, Manila, 1984.

13. **Prevett, P. F.,** The reduction of bruchid infestation of cowpeas by post-harvest storage methods, in *Annu. Rep. West African Stored Prod. Res. Unit,* Appendix 2, 1962.

14. **Giles, P. H.,** The storage of cereals by farm-

ers in Northern Nigeria, *Trop. Agric.,* 41, 197, 1964.

15. **Bugundu, L. M.,** The storage of farm products by farmers in my village, *Samaru Agric. Newslett.,* 12, 2, 1970.

16. **Lavie, D., Jain, M. K., and Shpan-Gabrielith, S. R.,** A locust phagorepellent from two *Melia* species, *Chem. Commun.,* 910, 1967.

17. **Butterworth, J. H. and Morgan, E. D.,** Isolation of a substance that suppresses feeding in locusts, *Chem. Commun.,* 23, 1968.

18. **Butterworth, J. H. and Morgan, E. D.,** Investigations of the locust feeding inhibition of the seeds of the neem tree, *Azadirachta indica, J. Insect Physiol.,* 17, 969, 1968.

19. **Morgan, E. D. and Thornton, M. D.,** Azadirachtin in the fruit of *Melia azedarach, Phytochemistry,* 12, 391, 1973.

20. **Nakanishi, K.,** Structure of the insect antifeedant azadirachtin, in *Recent Advances in Phytochemistry,* Vol. 9, Runeckles, V. C., Ed., Plenum Press, New York, 1975, 283.

21. **Zanno, P. R., Miura, I., Nakanishi, K., and Elder, D. L.,** Structure of the insect phagorepellent azadirachtin. Application of PRFT/CWD carbon-13 nuclear magnetic resonance, *J. Am. Chem. Soc.,* 97, 1975, 1975.

22. **Warthen, J. D., Jr., Redfern, R. E., Uebel, E. C., and Mills, G. D., Jr.,** Antifeedant for fall armyworm larvae from neem seeds, U.S. Department of Agriculture Research Results, ARR-NE-1, 1978.

23. **Warthen, J. D., Jr., Uebel, E. C., Dutky, S. R., Lusby, W. R., and Finegold, H.,** Adult housefly feeding deterrent from neem seeds, U.S. Department of Agriculture Research Results, ARR-NE 2, 1978.

24. **Kraus, W., Cramer, R., Bokel, M., and Sawitzki, G.,** New insect antifeedants from *Azadirachta indica* and *Melia azedarach,* in *Natual Pesticides from the Neem Tree (Azadirachta indica A. Juss),* Schmutterer, H., Ascher, K. R. S., and Rembold, H., Eds., GTZ Press, Eschborn, West Germany, 1981, 53.

25. **Dethier, V. G., Barton Browne, L., and Smith, C. N.,** The designation of chemicals in terms of the responses they elicit from insects, *J. Econ. Entomol.,* 53, 134, 1960.

26. **Pruthi, H. S.,** Report of the Imperial Entomologist, *Sci. Rep. Agric. Res. Inst.,* New Delhi, 123, 1937.

27. **Fry, J. S.,** Neem leaves as an insecticide, *Gold Coast Farmer,* 6, 190, 1938.

28. **Pruthi, H. S. and Singh, M.,** Pests of stored grain and their control, *Indian J. Agric. Sci.,* 18, 52, 1950.

29. **Krishnamurthy, B. and Rao, S. D.,** Entomol. Ser. Bull. No. 14, Agricultural College and Research Institute, Bangalore, India, 1950.

30. **Chopra, R. M., Bodhwar, P. L., and Ghosh,** S., *Poisonous Plants of India,* 2, Indian Council of Agricultural Research Publ., New Delhi, India, 1965.

31. **Borle, M. N.,** Exploration of the insecticidal efficacy of some indigenous plant products, Ph.D. thesis, Indian Agricultural Research Institute, New Delhi, India, 1966.

32. **Girish, G. K. and Jain, S. K.,** Studies on the efficacy of neem seed kernel powder against stored grain pests, *Bull. Grain Technol.,* 12, 226, 1974.

33. **Savitri, P. and Rao, C. S.,** Studies on the admixture of neem (*Azadirachta indica*) seed kernel powder with paddy in the control of important storage pests of paddy, *Andhra Agric. J.,* 23, 137, 1977.

34. **Akou-Edi, D.,** Effects of neem seed powder and oil on *Tribolium confusum* and *Sitophilus zeamais,* in *Natural Pesticides from the Neem Tree (Azadirachta indica A. Juss) and Other Tropical Plants* Schmutterer, H., Ascher, K. R. S., and Rembold, H., Eds., GTZ Press, Eschborn, West Germany, 1984, 445.

35. **Prakash, A., Pasalu, I. C., and Mathur, K. C.,** Save stored paddy from insects, *Indian Farming,* 30, 21, 1980.

36. **Jilani, G. and Su, H. C. F.,** Laboratory studies on several plant materials as insect repellants for protection of cereal grains, *J. Econ. Entomol.,* 76, 154, 1983.

37. **Laudani, H. and Swank, G. R.,** A laboratory apparatus for determining repellency of pyrethrum when applied to grain, *J. Econ. Entomol.,* 47, 1104, 1954.

38. **Jilani, G., Noorullah, and Ghiasuddin,** Studies on repellent properties of some indigenous plant materials against the red flour beetle, *Tribolium castaneum* (Hbst.), *Pak. Entomol.,* 6, 121, 1984.

39. **Jilani, G., Saxena, R. C., and Rueda, B. P.,** Repellent and growth inhibition effects of turmeric oil, sweetflag oil, neem oil, and 'Margosan O' against *Tribolium castaneum* (Coleoptera:Tenebrionidae), *J. Econ. Entomol.,* in press.

40. **Qadri, S. S. H.,** Some new indigenous plant repellents for storage pests, *Pesticides (India),* 7, 18, 1973.

41. **Pandey, N. D., Singh, S. R., and Tewari, G. C.,** Use of some plant powders, oils and extracts as protectants against pulse beetle, *Callosobruchus chinensis* Linn., *Indian J. Entomol.,* 38, 110, 1976.

42. **Jilani, G. and Malik, M. M.,** Studies on neem plants as repellent against stored grain pests, *Pak. J. Sci. Ind. Res.,* 16, 251, 1973.

43. **Laudani, H. D., Davis, D. F., and Swank, G. R.,** A laboratory method of evaluating the repellency of treated paper to stored-product insects, *Tech. Assoc. Pulp Paper Ind.,* 38, 336, 1955.

44. **McDonald, L. L., Guy, R. H., and Speirs, R. D.,** Preliminary evaluation of new candidate materials as toxicants, repellents, and attractants against stored-product insects-1, *U.S. Dep. Agric., Mark. Res. Rep.,* 882, 1970.

45. **Malik, M. M. and Naqvi, S. H. M.,** Screening of some indigenous plants as repellents or antifeedants for stored grain insects, *J. Stored Prod. Res.,* 20, 41, 1984.

46. **Saxena, R. C., Liquido, N. J., and Justo, H. D.,** Neem seed oil, a potential antifeedant for the control of the rice brown planthopper, *Nilaparvata lugens,* in *Natural Pesticides from the Neem Tree (Acadirachta indica A. Juss),* Schmutterer, H., Ascher, K. R. S., and Rembold, H., Eds., GTZ Press, Eschborn, West Germany, 1981, 171.

47. **Norris, D. M.,** Anti-feedant compounds, in *Chemistry of Plant Protection,* Haug, G. and Hoffmann, H., Eds., Springer, Berlin, 1986, 97.

48. **Jotwani, M. G. and Sircar, P.,** Neem seed as a protectant against stored grain pests infesting wheat seed, *Indian J. Entomol.,* 26, 161, 1965.

49. **Jotwani, M. G. and Sircar, P.,** Neem seed as a protectant against bruchid *Callosobruchus maculatus* (Fabricius) infesting some leguminous seeds, *Indian J. Entomol.,* 29, 21, 1967.

50. **Deshpande, A. D.,** Neem seed as a protectant against some of the foliage feeding insects, M.Sc. thesis, Indian Agricultural Research Institute, New Delhi, India, 1967.

51. **Chachoria, H. S., Chandratre, M. T., and Ketkar, C. M.,** Insecticidal trials against storage grain pests on maize seed with neem kernel powder, in *Report of Chief Plant Protection Officer,* Agric. Dept. Maharashtra, India, 1971.

52. **Saramma, P. U. and Verma, A. N.,** Efficacy of some plant products and magnesium carbonate as protectants of wheat seed against attack of *Trogoderma granarium, Bull. Grain Technol.,* 9, 207, 1971.

53. **Nair, M. R. G. K.,** Effect of neem seed kernel powder as protectant of stored grain, in *Utilization of Neem (Azadirachta indica Juss) and its Bye-Products,* Ketkar, C. M., Ed., 1976.

54. **Rajendran, R.,** Antifeeding studies against *Callosobruchus chinensis* Linn. (Bruchidae: Coleoptera) on red gram, green gram and *Spodoptera litura* F. (Noctuidae:Lepidoptera) and *Chirida bipunctata* (Cassididae:Coleoptera) on sweet potato, M.S. thesis, Tamil Nadu Agricultural University, Coimbatore, India, 1976.

55. **Subramaniam, T. R.,** Neem kernel powder against stored grain pests, in *Utilization of Neem (Azadirachta indica Juss) and its Bye-Products,* Ketkar, C. M., Ed., 1976.

56. **Siddig, S. A.,** Efficacy and persistence of powdered neem seeds for treatment of stored wheat against *Trogoderma granarium,* in *Natural Pesticides from the Neem Tree (Azadirachta indica A. Juss),* Schmutterer, H., Ascher, K. R. S., and Rembold, H., Eds., GTZ Press, Eschborn, West Germany, 1981, 251.

57. **Abdul Kareem, A.,** Neem as an antifeedant for certain phytophagous insects and a bruchid on pulses, in *Natural Pesticides from the Neem Tree (Azadirachta indica A. Juss),* Schmutterer, H., Ascher, K. R. S., and Rembold, H., Eds., GTZ Press, Eschborn, West Germany, 1981, 223.

58. **Schmutterer, H.,** Some properties of components of the neem tree (*Azadirachta indica*) and their use in pest control in developing countries, *Meded. Fac. Landbouwwet., Rijksuniv. Gent,* 46, 39, 1981.

59. **Sowunmi, O. E. and Akinnusi, O. A.,** Studies on the use of neem kernel in the control of stored cowpea beetle (*Callosobruchus maculatus* F.), *Trop. Grain Legume Bull.,* 27, 28, 1983.

60. **Pereira, J.,** The effectiveness of six vegetable oils as protectants of cowpeas and bambara groundnuts against infestation by *Callosobruchus maculatus* (F.) (Coleoptera: Bruchidae), *J. Stored Prod. Res.,* 19, 57, 1983.

61. **Devi, D. A. and Mohandas, N.,** Relative efficacy of some antifeedants and deterrents against insect pests of stored paddy, *Entomon,* 7, 261, 1982.

62. **Highland, H. A. and Wilson, R.,** Resistance of polymer films to penetration by lesser grain borer and description of a device for measuring resistance, *J. Econ. Entomol.,* 74, 67, 1981.

63. **Pereira, J. and Wohlgemuth, R.,** Neem (*Azadirachta indica* A. Juss) of West African origin as a protectant of stored maize, *Z. Angew. Entomol.,* 94, 208, 1982.

64. **Lalkani, G. K. and Patel, N. G.,** Non-pesticidal control of *Ephestia cautella* Walker, *Pesticides (India),* 19, 46, 1985.

65. **Jilani, G. and Haq, H. S.,** Studies on some indigenous plant materials as grain protectants against insect pests of stored grains, *Pak. Entomol.,* 6, 24, 1984.

66. **Prakash, A., Pasalu, I. C., and Mathur, K. C.,** Evaluation of plant products as grain protectants against insect pests of stored paddy, *Bull. Grain Technol.,* 18, 25, 1980.

67. **Prakash, A., Pasalu, I. C., and Mathur, K. C.,** Plant products for management of stored grain insect pests, *Bull. Grain Technol.,* 19, 213, 1981.

68. **Verma, S. P., Singh, B., and Singh, Y. P.,**

Studies on the comparative efficacy of certain grain protectants against *Sitotroga cerealella* Olivier, *Bull. Grain Technol.*, 21, 37, 1983.

69. **Jhansi Rani, B.,** Studies on the biological efficacy of deoiled neem *(Azadirachta indica* A. Juss) kernel against a few insects, M.S. thesis, Indian Agricultural Research Institute, New Delhi, India, 1984.

70. **Pathak, P. H. and Krishna, S. S.,** Neem seed oil, a capable ingredient to check rice moth reproduction (Lepidoptera:Galleriidae), *Z. Angew. Entomol.*, 100, 33, 1985.

71. **Sharma, G. K., Czoppelt, Ch., and Rembold, H.,** Further evidence of insect growth disruption by neem seed fractions, *Z. Angew. Entomol.*, 90, 439, 1980.

72. **Rembold, H., Sharma, G. K., and Czoppelt, Ch.,** Growth-regulating activities of azadirachtin in two holometabolous insects, in *Natural Pesticides from the Neem Tree (Azadirachta indica A. Juss),* Schmutterer, H., Ascher, K. R. S., and Rembold, H., Eds., GTZ Press, Eschborn, West Germany, 1981, 121.

73. **Singh, K. N. and Srivastava, P. K.,** Neem seed powder as a protectant against stored grain insect pests, *Bull. Grain Technol.*, 19, 127, 1980.

74. **Qadri, S. S. H., Rao, B. B., and Brahmanand, B.,** Effect of combining some indigenous plant seed extracts against household insects, *Pesticides (India)*, 11, 21, 1977.

75. **Paul, C. F., Dixit, R. S., and Agarwal, P. N.,** Evaluation of the insecticidal properties of the seed oil and leaf extract of the common Indian neem, *Azadirachta indica* Linn., *Sci. Cult.*, 29, 412, 1963.

76. **Atwal, A. S. and Sandhu, G. S.,** Preliminary studies on the efficacy of some vegetable and inert dusts as grain protectants, *J. Res.*, 7, 52, 1970.

77. **Jilani, G., Noorullah, and Khan, M. I.,** Use of local plant materials for protection of stored grains against insect pests, 2nd Annu. Rep. IFS Project, Pest Management Research Institute, Pakistan Agricultural Research Council, Karachi, Pakaistan, 1986.

78. **Parmar, B. S. and Datta, S.,** Neem oil as a synergist for insecticides, *Neem Newslett.*, 3 (1), 3, 1986.

79. **Bains, S. S., Battu, G. S., and Atwal, A. S.,** Effect of powdered neem *(Azadirachta indica* A. Juss) material on the diapause in larvae and population build-up of *Trogoderma granarium* Everts, infesting stored wheat, *Indian J. Plant Prot.*, 4, 192, 1977.

80. **Zhang, X. and Zhao, S. H.,** Experiments on some substances from plants for the control of rice weevils, *J. Grain Storage (Liangshi Chucang)*, 1, 1, 1983.

81. **Ivbijaro, M. F.,** Toxicity of neem seed, *Azadirachta indica* A. Juss to *Sitophilus oryzae* (L.) in stored maize, *Prot. Ecol.*, 5, 353, 1983.

82. **Bowry, S. K., Pandey, N. D., and Tripathi, R. A.,** Evaluation of certain oil seed cake powders as grain protectant against *Sitophilus oryzae* Linnaeus, *Indian J. Entomol.*, 46, 196, 1986.

83. **Rout, G.,** Comparative efficacy of neem seed powder and some common plant powder admixtures against *Sitophilus oryzae* (Linn.), *Neem Newslett.*, 3 (2), 13, 1986.

84. **Qadri, S. S. H.,** Behavioral and physiological approaches to control insect pests. in *Behavioral and Physiological Approaches in Pest Management,* Regupathy, A. and Jayaraj, S., Eds., Tamil Nadu Agricultural University, Coimbatore, India, 1985, 47.

85. **Yadav, T. D.,** Studies on the isecticidal treatments against bruchids *Callosobruchus maculatus* (Fab.) and *C. chinensis* (Linn.) damaging stored leguminous seeds, Ph.D. thesis, Agra University, Agra, India, 1973.

86. **Yadav, T. D.,** Efficacy of neem *(Azadirachta indica* A. Juss) kernel powder as seed treatment against pulse beetles, *Neem Newslett.*, 1 (2), 13, 1984.

87. **Ivbijaro, M. F.,** Preservation of cowpea *Vigna unguiculata* (L.) Walp, with the neem seed, *Azadirachta indica* A. Juss, *Prot. Ecol.*, 5, 177, 1983.

88. **Sangappa, H. K.,** Effectiveness of oils as surface protectants against the bruchid, *Callosobruchus chinensis* Linnaeus infestation on red gram, *Mysore J. Agric. Sci.*, 11, 391, 1977.

89. **Das, G. P.,** Pesticidal efficacy of some indigenous plant oils against the pulse beetle, *Callosobruchus chinensis* (Coleoptera, Bruchiidae), *Bangladesh J. Zool.*, 14, 15, 1986.

90. **Ali, S. I., Singh, O. P., and Misra, U. S.,** Effectiveness of plant oils against pulse beetles *Callosobruchus chinensis* Linn., *Indian J. Entomol.*, 45, 6, 1983.

91. **Yadav, T. D.,** Antiovipositional and ovicidal toxicity of neem *(Azadirachta indica* A. Juss) oil against three species of *Callosobruchus*, *Neem Newslett.*, 2 (1), 5, 1985.

92. **Zehrer, W.,** The effect of the traditional preservatives used in Northern Togo and of neem oil for control of storage pests, in *Natural Pesticides from the Neem Tree (Azadirachta indica A. Juss) and other Tropical Plants,* Schmutterer, H. and Ascher, K. R. S., Eds., GTZ Press, Eschborn, West Germany, 1984, 453.

93. **Ketkar, C. M.,** Use of tree-borne non-edible oils like neem, karanja, undi, sal and kusum as surface protectants for stored cowpea seed against *Callosobruchus maculatus* and for green gram against *C. chinensis*, in *Proc. 3rd*

Int. Neem Conf., Schmutterer, H., and Ascher, K. R. S., Eds., Nairobi, Kenya, GTZ Press, Eschborn, West Germany, 1987, 543.

94. **Ambika, B., Abraham, C. C., and Nalinakumari, T.,** Effect of neem leaf extract and two JH analogues on the development of *Callosobruchus chinensis* Linn. (Coleoptera: Bruchidae), *Agric. Res. J. Kerala,* 19, 72, 1981.

95. **Muda, A. R.,** Utilization of neem as a pest control agent for stored paddy, in *Health and Ecology in Grain Post-Harvest Technology,* Sample, R. L. and Frio, A. S., Eds., ASEAN Crops Post-Harvest Programme and ASEAN Food Handling Bureau, Manila, 1984, 117.

96. **Jilani, G.,** Evaluation of some botanical products as grain protectants against insect pests at farm level in Pakistan, in *17th Annu. Convention,* PCCP, Iloilo, Philippines, 1986 (Abstr.).

97. **Jilani, G. and Amir, P.,** Economics of neem in reducing wheat storage losses: policy implications, Tech. Bull. 2, SEARCA, Philippines, 1987.

98. **Jacobson, M.,** Neem research in the U.S. Department of Agriculture: chemical, biological and cultural aspects in *Natural Pesticides from the Neem Tree (Azadirachta indica A. Juss),* Schmutterer, H., Ascher, K. R. S., and Rembold, H., Eds., GTZ Press, Eschborn, West Germany, 1981, 33.

99. **Qadri, S. S. H., Usha, G., and Jabeen, K.,** Sub-acute dermal toxicity of Neemrich-100 (tech.) to rats, *Int. Pest Control,* 26, 18, 1984.

100. **Saxena, R. C., Waldbauer, G. P., Liquido, N. J., and Puma, B. C.,** Effects of neem seed oil on the rice leaffolder, *Cnaphalocrocis medinalis,* in *Natural Pesticides from the Neem Tree (Azadirachta indica A. Juss),* Schmutterer, H., Ascher, K. R. S., and Rembold, H., Eds., GTZ Press, Eschborn, West Germany, 1981, 189.

7

The Effects of Neem on Insects Affecting Man and Animal

K. R. S. Ascher and J. Meisner
Department of Toxicology
Agricultural Research Organization (ARO)
The Volcani Center
Bet Dagan, Israel

TABLE OF CONTENTS

I. INTRODUCTION

The neem tree or Indian lilac, *Azadirachta indica* A. Juss (= *Antelaea azadirachta* L.), also known under many local names such as Margosa (Portuguese), Mawarobaini (Swahili), etc., is an evergreen tree growing in the Indian subcontinent, Southeast Asia, and many African countries. It is very hardy, surviving also under unfavorable climatic conditions, and demonstrating resistance to attack by many insect pest species and other pest organisms such as nematodes. The botanical aspects of neem have been reviewed in recent years by Howaldt[1] and Anon. (National Academy of Sciences).[2] Apart from its wood, which can be used for the production of furniture, the most valued part of the neem tree is its seed, the oil of which (up to 40% content of the seed) is used in developing countries as lamp fuel and by the pharmaceutical industry. Moreover, the so-called neem cake (remaining after the extrusion of the oil) is used as fertilizer. The neem seed and other parts of the tree contain insecticidal tetranortriterpenoids such as azadirachtin,[3] meliantriol,[4] and salannin,[5] which act in insects as antifeedants, as insect growth and development regulators, or even as toxicants. The most potent compound of the three is azadirachtin.*

Azadirachtin was first isolated from *Azadirachta indica*[3] and later from the Persian lilac, *Melia azedarach* L.[8] by Morgan and co-workers. Structures were proposed by Zanno et al.[9] in 1975 and by the groups of Ley and Morgan[10] in 1985, but the final, correct structure was advanced recently (1985) by Kraus et al.[11] The insecticidal properties of different neem formulations and of the three pure compounds mentioned above have been reviewed by several workers, e.g., by Ketkar,[12] Warthen,[13] Jain,[14] and Jacobson.[15] Much material is presented in the *Proceedings of the 1st, 2nd, and 3rd Neem Conferences* which were held in West Germany (1980 and 1983) and in Kenya (1986). Most of the publications on the insecticidal effects of neem deal with insects of agricultural importance, and stored-product pests. Much less work has been done on insects of medical and hygienic importance, and on household pests, which are the subject of the present review.

II. HOUSE FLIES AND BLOW FLIES

A. HOUSE FLIES

Quadri and Rao[16] investigated the toxicity of neem seed extract to house flies. Adults of the Indian house fly, *Musca domestica nebulo* F., 3 to 5 d old, were exposed to a dried ether extract of neem seeds in the form of residues from acetone in 10-cm-diameter petri dishes. The LC_{50} of neem was 170 μg/dish vs. 7 μg/dish for DDT. A 1:5 mixture of custard apple (*Annona squamosa* L.) and neem seed extracts was as toxic (~6 μg/dish) as DDT to *M. dom. nebulo*. Other mixing ratios (1:4, 1:6, 1:8) were less active. These authors[17] also examined the synergism exerted by neem seed extract to pyrethrum in adults of the Indian house fly. Neem synergizes a 30% pyrethrum extract (oleoresin concentrate) when added as surplus at a ratio of 20:1. The LC_{50} of neem in these experiments was 182 μg/dish, of oleoresin 5 μg/dish, and of the mixture ~0.8 μg/dish.

Gill[18] found that feeding of adult house flies, *M. domestica* L., for 3 d after emergence on a ground mixture of 10% neem kernel dust with sucrose and then transferring them to untreated milk soaked in cotton wool did not affect the fecundity, egg-fertility, or emergence of the resultant adults. Thus neem had no sterilizing effect on the adult house fly.

---

* Forster[6] and Rembold[7] state that there are three azadirachtins: A, B, and C.

Naqvi[19] topically applied a petroleum ether solution of the neutral fraction of ripe neem fruits (mixture of the triterpenoids) to adult house flies (see also Saeed and Naqvi[20]). On the basis of mortality readings taken after 24 h, the LD_{50} was 1.4 µg/fly. Cholinesterase, acid phosphatase, and alkaline phosphatase of insects treated at the LD_{50} level were inhibited by 40 to 50% as compared with untreated insects. Chavan[21] tested a fraction (NP-2) obtained from petroleum ether neem leaf extracts (see below) against house flies in the Peet-Grady Chamber, as a 2% formulation; 82% mortality was obtained.

Gill[18] mixed neem kernel dust in the range of 1 to 10% into the larval medium of *M. domestica* and infested each replicate with 50 2-d-old larvae. No pupae developed at the 10, 5, and 2.5% neem levels in the food. At 1% neem, 2% of the larvae pupated and 1% adults were obtained; pupation rate in the control being 79% and adult emergence, 51%. The larvae in the 10 and 5% treatments died within 1 week and those in the 2.5 and 1% treatments grew very slowly. The larval duration of the few larvae reaching pupation at 1% was prolonged up to 25 d, vs. only 7 d in the control. There was, of course, a marked reduction of the growth rate of treated larvae. Gill believed that neem dust inhibited larval feeding and that the larvae died from starvation, although he did not exclude hormonal control of growth. The toxicity was somewhat lower in analogous experiments conducted by Ahmed et al.,[22] who incorporated powdered neem seed kernel at high concentrations into a *M. domestica* culture medium; this proved to be only slightly toxic to third instar larvae (31% kill) at 40% in the medium. Pupal mortality, however, ranged from 90% for 40 and 30% neem seed kernel, to 76 and 57% with 20 and 10% incorporated, respectively.

Perera[23] mentioned in 1941 that neem oil is widely used in India in the treatment of livestock. He stated that neem bitters and the characteristic smell render it an efficient and cheap wound dressing, by repelling flies. Indeed, the latter property was confirmed by Warthen et al.[5] Jacobson[24] reported that housefly adults were repelled by neem in USDA tests. In fact, Warthen et al.[5] at the USDA investigated the repellence of fractions and pure compounds from neem seeds for *M. domestica*. For the bioassays, the test materials in acetone were added to sucrose and the solvent was removed *via* a rotary evaporator. Treated and untreated sucrose (control) samples on watch glasses in petri dishes were exposed for 3 min to 3- to 10-d-old flies. The flies were then driven away and the count was repeated. Bioassay-monitored fractionation of products obtained from ground neem seed extracted with 95% ethanol by partition of the extract between methanol and hexane, chromatography of the methanol-soluble portion on a Florex® RVM column, subsequent separation of an active fraction on a Florisil® column, followed by chromatography on a Bio-Sil® HA column, and finally HPLC, yielded a potent house fly feeding deterrent, which was shown to be salannin. Warthen et al. compared the deterrent effect of salannin with that of diethyltoluamide (Deet®); Zoecon® 0759 (15% MGK 326 = dipropyl pyridine-2,5-dicarboxylate); Zoecon® 0760 [15% SC Johnson R-69 = 3-acetyl-2-(2,6-dimethyl-5-heptenyl)oxazolidine]; limonene; cedrene; and other repellents. The results were expressed as percent flies settling on treated/untreated sugar. The values for 0.5, 0.25, and 0.1% salannin-treated sugar were 0, 10, and 0%, respectively. Deet® gave 0% at 0.5% but 54% at 0.1% (no results for 0.25%). Zoecon® 0759 and 0760 gave 100 % deterrence even at 0.1% incorporated substance. Cedrene and limonene were much less active (100% deterrence only at 1%). Gill[18] offered neem kernel water extract, alcohol extract, or azadirachtin at the extremely low concentration of 100 µg/l (0.1 ppm) in a 1 *M* sucrose solution to adult house flies for 3 d. The flies fed normally in all the treatments. This was proved by starving them for a further 3 d, which caused 100% mortality. We believe that (a) this concentration was too low to obtain any deterrent effect and (b) the neem or azadirachtin dilutions, and the sucrose, should have been offered separately, e.g., neem derivatives in water and sucrose in the solid state, since sucrose in solution with neem may overcome the deterrent effect of the latter. Emerging house flies were not deterred from feeding on a ground mixture of 10% neem kernel dust with sucrose.[18]

B. SWARMING HOUSE FLIES

Gaaboub and Hayes[25] exposed third instar larvae of the face fly, *Musca autumnalis* De Geer, to filter paper disks (9-cm diameter) treated with 5 ml of solutions of various concentrations of azadirachtin in acetone, ranging from 10 pg to 100 μg/ml, for either 10 or 20 min. Development to adults was inhibited by very low concentrations, with 10 pg causing 20 and 22%, and 0.1 μg causing 83 and 98% inhibition of adult emergence, at the 10- and 20-min exposure times, respectively. Adult inhibition concentrations were calculated and the IC_{50}s were found to be very low: 240 and 39 pg for the 10- and 20-min exposure times, respectively. Deleterious effects on pupae (undeveloped individuals) and adults (incomplete development, attachment to puparia, or inability to fly) were dose-dependent. With higher concentrations mortality set in already in the larval stage and during the development from larva to pupa. No adults were obtained at 1.0 μg/ml or higher doses, with mortality setting in in the larval stage or in the pupa. Very low concentrations of azadirachtin (10 pg to 0.1 μg/ml) caused a significant reduction in pupal weight and dimensions, adult weight, wing dimensions (of males only), or the interocular distance. Gaaboub and Hayes,[26] moreover, found that adults developing from larvae that had survived the 20-min exposure to 10 and 30 pg/ml (i.e., the approximate IC_{25} and IC_{50} for this exposure time), had reduced fecundity and egg hatch. When both males and females were survivors from treated larvae (T♂♂ × T♀♀) reduction in fecundity was 67 and 86%, and reduction in egg hatch was 37 and 73% for the two concentrations, respectively. When only one of the sexes was treated, decrease in fecundity was ~30% at the lower concentration and 47% at the higher concentration, irrespective of whether males or females were treated. The decrease in hatchability was, at the lower concentration, 32% for T♂♂ × UT♀♀ and 12% for UT♂♂ × T♀♀ and at the higher concentration, 60% for T♂♂ × UT♀♀ and 70% for UT♂♂ × T♀♀. However, since fecundity is presented in the paper of Gaaboub and Hayes as average number of eggs/lifetime of female (and not as number of eggs/female-days), and since the longevity of treated males and females was strongly reduced, our calculations show that the reduced fecundity is due to the shortened life-span of the treated flies. In fact, the differences in fecundity disappear when the fecundity is recalculated as eggs/female-days and only the decrease of hatchability remains a valid phenomenon.

C. BLOW FLIES

Hobson[27] sprayed or dipped sheep with repellent oils and then tested the response to the oviposition of *Lucilia sericata* Meig. at different intervals after the treatment. The test was conducted by inserting into the fleece a piece of cotton wool soaked with a blow fly attractant solution containing ammonium carbonate and indole. A 3% emulsion of neem oil sprayed over the back of the animal deterred blow flies from ovipositing for 8 d, which was longer than with some other oils (*Tagetes*, rape and whale oil, and kerosene) but less effective than olive, cotton seed, castor and soybean oil, and oleic acid. Rice et al.[28] investigated the oviposition-deterring effect of crude neem oil and an azadirachtin-enriched, 30% methanolic, dewaxed neem extract (AZT) on gravid females of the sheep blow fly *Lucilia cuprina* (Wied.). The tests were conducted with artificial oviposition substrates consisting of sheep plasma-impregnated oviposition pads. Neem formulations proved to be powerful antioviposition agents for *L. cuprina*. AZT provided total repellence at 0.02%, whereas undiluted neem oil gives 91% deterrence. AZT is thus much more potent than neem oil, probably because of its enhanced azadirachtin content.

Bidmon et al.[29] investigated biological effects of azadirachtin in blow fly larvae. Injection of azadirachtin into larvae of the blue blow fly, *Calliphora vicina* R.-D., led to several dose- and stage-dependent biological effects, such as (1) a delay of pupariation when injected in the first half of the last larval instar (no effect when injected in the late instar); (2) reduced pupal weight; and (3) dose-dependent inhibition of adult emergence. The adults that succeeded in emerging from azadirachtin-treated larvae were smaller and demonstrated different types of malformations. These investigators also determined the effect of azadirachtin on the ecdysteroid titer of larvae and pupae. The larval peak of ecdysteroid which triggers puparium formation is delayed in azadirachtin-treated larvae, and it is lower and lasts longer. High doses of azadirachtin also delay the large pupal peak of ecdysteroids, and the decline of ecdysteroid titer is slower in pupae after larval treatment with azadirachtin. Since the eclosion hormone is controlled by ecdysteroids, Bidmon et al. believe that the cause of inhibition of adult eclosion by azadirachtin is the persistent high concentration of ecdysteroids in pupae. Of interest was the effect of azadirachtin on the production of ecdysone biosynthesis and release: azadirachtin inhibited the release of ecdysone from the isolated brain/ring gland complexes *in vitro*. Moreover, the rate of ecdysone metabolism (hydroxylation to 20-hydroxyecdysone) in the isolated fat body was decreased by azadirachtin *in vitro* in a dose-dependent manner. Azadirachtin thus has a number of effects on different parts of the ecdysteroid system.

III. BLOODSUCKING FLIES

A. STABLE FLIES

Radwanski[30] states that larvae of *Musca, Stomoxys,* and *Anopheles* failed to develop on a diet containing neem kernels.

Gill[18] placed batches of *Stomoxys calcitrans* (L.) eggs on filter paper moistened with 1% aqueous neem seed kernel extract (NSKE), 1% aqueous fruit extract, or 0.01% azadirachtin. Only the last material reduced egg hatch to 7 vs. 88% in control, whereas in the 1% aqueous NSKE and fruit extract hatch was 67 and 84%, respectively. Thus an ovicidal effect was shown only by azadirachtin proper. Some of the eggs treated with azadirachtin exhibited longitudinal fissures, partly exposing the larvae. In further experiments conducted by Gill, neem kernel dust was incorporated in the range of 0.1 to 10% in a standard *Stomoxys* larval diet and 15 3- to 4-d-old larvae of *S. calcitrans* were introduced per replicate. In the 10% treatment all larvae died within 2 d; in the 5% treatment 91% died after 2 d and the rest after 7 d. At 1%, mortality did not exceed 40% after 2 and 7 d and reached 65% after 10 d; results with 0.1% were no different from control. Neem affected the larval growth, with reduced larval weight, prolonged larval duration, delayed pupation, and reduction in adult emergence.

B. MOSQUITOES
1. *Aedes spp.*

Schmutterer and Zebitz[31] assayed the response of young fourth instar *Aedes aegypti* (L.) larvae to methanolic NSKEs obtained from trees of various geographic provenience: southern and northern Togo, central Sudan, southern India, and upper Burma. The larvae were exposed continuously to different concentrations of the neem extracts in the breeding water. There was no direct toxicity to the larvae, but rather IGR effects. The majority of the lethally affected insects died as white, slightly or normally melanized pupae, with adult structures visible through the skin, about 6 to 7 d after the beginning of the experiment. In a small proportion of the dead pupae protruded mouth parts and deformed wing sheaths were evident. The LC_{50}s of the various extracts did not differ greatly, ranging from 55 to 65 mg/l. Other members of Schmutterer's team (Ermel et al.[32]) found that neem extracts from these seeds of different provenience (Togo, Sudan, India, and Burma) had an IGR efficacy in *A. aegypti* which was not necessarily correlated with the azadirachtin contents of the extracts.*

Zebitz[33] extracted neem seed kernels either with water (aqueous neem seed kernel extract, ANSKE), with methanol (MeOH-E), or with various solvents following a primary extraction with petroleum ether (Vorreinigung = VR) to remove fats; with methanol (MeOH-VR-E); butanol (BuOH-VR-E); methylethyl ketone (MEK-VR-E); or methyl-*tert*-butyl ether (MTB-VR-E). Two further extracts were prepared with an azeotropic mixture (AZT) of MeOH and MTB with or without primary removal of fats with petroleum ether (VR), and in both cases with separation of the remaining fat by solidification by freezing (Kalt = K) and subsequent separation from the extracts by filtration. Consequently, these two extracts were designated AZT-K-E and AZT-VR-K-E. All these eight extracts may be classified as variously prepared crude extracts. A further two azadirachtin-enriched liquid/liquid extracts (MTB/H_2O-VR-E and MTB/H_2O-NR-K-E, where NR = Nachreinigung, further postextraction solvent/solvent separation of fat with petroleum ether and final filtration after freezing out the remaining fat) were prepared. Zebitz tested the effect of these extracts on first and fourth instar larvae of *A. aegypti* continuously exposed to the treated water. In treatment of fourth instar larvae, the mortality with the 20 mg/l treatment was highest with the MTB/H_2O-VR-E (100%) followed by MTB-VR-E (77%), MEK-VR-E (69%), BuOH-VR-E (49%) and AZT-K-E (25%). The two methanol extracts (MeOH-E and MeOH-VR-E) were inactive at this concentration. LC_{50}s were calculated for ANSKE (78 mg/l), AZT-VR-K-E (18 mg/l), and MTB/H_2O-NR-K-E (6 mg/l). These results show that the aqueous extract was less active than several extracts prepared from organic solvents; e.g., BuOH-VR-E, MEK-VR-E, MTB-VR-E, MTB/H_2O-VR-E, MTB/H_2O-NR-K-E, and AZT-VR-K-E were more active than ANSKE. Most of the insects killed by ANSKE, AZT-VR-K-E, and MTB-VR-E died as pupae with adult forms visible, but MTB/H_2O-NR-K-E killed earlier stages. With continuous exposure from first instar larvae until adult emergence, approximately 90% mortality was brought about by 16 mg/l ANSKE, 6 mg/l AZT-VR-K-E, and 2 mg/l MTB/H_2O-NR-K-E. AZT-VR-K-E had the strongest growth-retarding effect among the three extracts; this was ascribed to its high salannin content; salannin was not detected in MTB/H_2O-NR-K-E and was present only in trace amounts in ANSKE. The mortality in the larval stage only in these experiments was 18, 26, and 39% for MTB/H_2O-NR-K-E, AZT-VR-K-E, and ANSKE, respectively. This seems to indicate that the larvicidal effect of the extracts increased with rising polarity of the extraction solvents.

The effects of some of these extracts were investigated in further mosquito species and compared with *A. aegypti*. Thus Zebitz[34,35] assayed ANSKE, AZT-VR-K-E, and MTB/H_2O-NR-K-E against young fourth instar larvae of *A. aegypti, A. togoi* (Theobald), *Culex quinquefasciatus* Say, and *Anopheles stephensi* List. He found conspicuous differences in the susceptibility of the different species against the extracts. Judging from the LC_{50} values (in parentheses) of AZT-VR-K-E, susceptibility decreased in the following order: *Aedes togoi* = *Anopheles stephensi* > *C. quinquefasciatus* > *Aedes aegypti*; the respective LC_{50}s for AZT-VR-K-E were 1.2, 1.2, 4.9, and 18.1 mg/l. This order of activity against these mosquito species remains the same also with the other extracts, although ANSKE was less active than AZT-VR-K-E, whereas MTB/H_2O-K-NR-E was more active (LC_{50} of the latter against *A. togoi* was 0.37 mg/l). The biological activity

---

* Statement found in the abstract of the paper only.

of pure azadirachtin against *A. togoi* (LC_{50}, 0.22 mg/l) was only slightly less than twice that of MTB/H_2O-NR-K-E.

More than 90% of the injured insects died in the pupal stage. The teratologic symptoms appear to be the same as those caused by methoprene. Yet it seems that the NSKEs and azadirachtin have a mode of action different from that of the JHA since it is considered that azadirachtin interferes directly or indirectly with the ecdysone titer of the insects. Also, the slope of the log concentration-probit line of azadirachtin is completely different from that of methoprene, another indication of two different modes of action. Moreover, whereas freshly molted fourth instar mosquito larvae are the most susceptible phase against azadirachtin-containing NSKEs, methoprene is the most toxic against fourth instars shortly before pupation. Finally, Zebitz found that extracts from neem leaves which contain no azadirachtin bring about the same effect as the pure substance and NSKEs. It is assumed that leaves contain related substances with a mode of action equivalent to or resembling that of azadirachtin.

Continuous exposure of larvae of *A. aegypti* starting from the first instar, to NSKE-treated water, led to a 4-d delay in adult emergence.

Zebitz[35] treated water in mosquito rearing bowls with AZT-VR-K-E or an azadirachtin-enriched neem oil (containing 400 μg azadirachtin/g) for studying oviposition deterrency with *C. quinquefasciatus*. Filter paper moistened with treated water was used in analogous experiments with *A. aegypti*. Gravid females of *A. aegypti* were not deterred from oviposition on filter paper treated with either neem formulation, whereas *C. quniquefasciatus* demonstrated considerable oviposition deterrency to neem oil but not to AZT-VR-K-E. In none of the experiments was the larval hatch from the egg rafts reduced, except for *C. quinquefasciatus* eggs in AZT-VR-K-E-treated water.

In biting deterrency tests, the biting rate of female *A. aegypti* increased with time and reached that of control nearly 90 min after application of neem oil to hands. Only the biting rates counted immediately and 30 min after administration of the oil to the human skin differed significantly from control.[35]

Some work was also done with a combination of neem with a microbial pesticide. Hellpap and Zebitz[36] assayed mixtures of AZT-VR-K with a *Bacillus thuringiensis* var. *israelensis* (serotype H-14) preparation against 0- to 16-h-old and 1- to 4-d-old fourth instar larvae of *A. togoi*. AZT-VR-K concentrations in the breeding water ranged from 0.5 to 1.5 mg/l for the younger, and from 1.0 to 2.5 mg/l for the older fourth instars; the *B. thuringiensis* preparation concentrations ranged from 0.01 to 0.1 mg/l for both age groups. The effect of the two pesticides was additive, with a few exceptions of synergistic effect. The kill obtained by AZT-VR-K alone in the young (16 h) fourth instar larvae was only 4% with the 0.5 mg/l concentration, 36% with 1 mg/l, and 65% with 1.5 mg/l. The preparation was less active against slightly older larvae (1- to 4-d-old fourth instars): 1 mg/l — zero mortality; 2.5 mg/l — 41% mortality.

To examine the activity of some specific compounds isolated from neem, Naqvi[19] exposed fourth instar larvae of *A. aegypti* (see also Naqvi and Ahmed[37]) to various concentrations of the neutral fraction of neem winter leaves (I) containing nimocimolide (II) and isonimocimolide (III) as major components, as well as to II and III alone. The LC_{50} of I, II, and III was 0.6 to 0.7 mg/l. Treated larvae produced adults with deformed wings and legs, and a shrunken abdomen. Larval/pupal intermediates were also observed.

Jacobson[24] mentioned (personal communication from C. E. Schreck and D. E. Weidhaas) *A. aegypti* adults as being repelled by neem in USDA tests; azadirachtin was more effective than salannin. Cloth treated with 1 mg/cm² azadirachtin and placed on the forearm of human test subjects exposed for 1 min in cages of *A. aegypti*, protected the subject from bites through clothes, but 0.2 mg/cm² was much less effective. Detailed data from the Insects Affecting Man and Animal Research Laboratory at Gainesville, Fla. (M. Jacobson, personal communication, and Jacobson[24]) showed that azadirachtin in mosquito (*A. aegypti*) cloth repellency tests was less active than Deet® and DMP (dimethyl phthalate). A fresh 15-min air-dried treatment with 0.2 mg azadirachtin/cm² allowed four bites during a 1-min exposure (fresh treatments of the Deet® and the DMP standards are effective at the minimum rate of 0.016 mg/cm²). A 1-d-old azadirachtin treatment allowed 14 bites within 1 min. Salannin was even less active: a fresh, 15-min air-dried treatment of 1 mg/cm² allowed 25 bites during the 1-min exposure and the 0.5 mg/cm² treatment allowed 27 bites; the same dosages of the Deet® and DMP standards were 100% effective (no bites), both as fresh and 24-h-old treatments.

Mwangi and Rembold[38] investigated the toxicity of the fruit extract of a further *Melia* sp., *M. volkensii*, against *A. aegypti*. Fresh ripe fruits from this tree from Kenya were dried at 40°C to constant weight and ground to a 1-mm-mesh powder; this drying temperature was found to be optimal. The powder was extracted with methanol, the extract was lyophilized, dissolved in acetone, and fractionated on a silica gel column by elution with different solvents. The fractions obtained were dissolved in absolute ethanol and bioassayed with *A. aegypti* second instar larvae by continuous exposure until the pupal stage. Most of the larvicidal activity was contained in a 1:1 hexane:ethyl acetate fraction. The authors found no azadirachtin in the active fraction and believe that the active compound(s) must be more active than azadirachtin.

2. *Culex* spp.

Singh[39] investigated the effect of some NSKEs on second instar larvae of *Culex fatigans* Wied. Whereas an aqueous and an alcohol extract gave inconsistent results, consistent results were obtained with the aqueous extract of powdered kernel previously deoiled with petroleum ether. After 2, 4, 6, and 8 d, 100% mortality in the 500, 250, 125, and 62.5 mg/l concentrations, respectively, was obtained; with 31.2 mg/l 85% mortality was obtained after 12 d, the survivors reaching the adult stage.

Chavan et al.[40] (see also Chavan[21]) extracted dried neem leaf powder with petroleum ether, ether, chloroform, and ethanol. Within 24 and 48 h, 1% petroleum ether extract, 0.2% ether extract,* and 0.5% pyrethrum extract* gave 100% mortality of larvae of *C. pipiens fatigans* in treated water. The residual activity then declined, but was still 50 to 60% after 6 d. Work was continued only with the petroleum ether extract, because the yield of the ether extract was very low. The petroleum ether extract was further fractionated by neutral alumina column chromatography, using successively petroleum ether, benzene, 1:1 benzene:methanol, and methanol as eluting solvents. The benzene eluate was somewhat less active than the petroleum ether extract, allowing the development of 10% larvae to dead pupae, but not of live adults. Fraction NP-2 obtained from the benzene eluate by further purification produced 100% mortality at concentrations ranging from 100 to 10 mg/l. As regards the residual effect, 100 mg/l gave 100% kill for 8 d, 50 and 25 mg/l for 6 d, and 10 mg/l for 5 d.

Chavan[21] also tested fraction NP-2 as a 2% formulation in a Peet-Grady chamber against adult *C. p. fatigans* mosquitoes; the mortality achieved was 73%.

Attri and Prasad[41] (cited also by Jotwani and Srivastava[42]) found that neem oil extractive, a waste product from neem oil refining (about 10% of the crude oil), was an effective mosquito larvicide. Its composition is stated to be volatile matter or alcohol, ~2.0%; crude bitters, ~23%; fatty ballast, ~53%; and sulfur compounds with a strong odor, ~22%. An emulsifiable concentrate was prepared by mixing the extractive with

---

* These concentrations are according to Chavan.[21] According to Chavan et al.,[40] the figures were 0.5% ether extract and 0.2% pyrethrum extract.

10% (by volume) of Tween® 80. Aqueous dilutions of this concentrate were tested in the laboratory against first instar larvae of *C. fatigans* in beakers. Mortalaity was recorded for 7 d. Down to a concentration of 0.01%, none of the larvae survived; 100% of the larvae were killed at 0.05 and 0.04% on the first day, and at 0.03 to 0.01% by the seventh day. At 0.005% only 60% of the larvae died after 7 d, but none of the survivors was able to emerge as an adult; 0.001% killed only 40% of the larvae, whereas the rest pupated and emerged as normal adults. Although 0.005% was innocuous to the insectivorous fish *Gambusia* sp., 0.01% was toxic (40% kill) and 0.04% was highly toxic (100% mortality) within 24 h. This indicates that neem oil extractive is a serious hazard to insectivorous fish and probably other species of fish and precludes its use in fish ponds. Toxicity to tadpoles was also high; with 0.02%, 30% kill within 4 d; with 0.04%, 80% after 1 d and 100% after 2 d. The authors point out the potential of the formulation, at least in cities as a replacement for petroleum products. This waste product is available in amounts totaling 8300 tons/year in India, can be formulated easily and, unlike other neem products, is quite stable. The same authors[43] found neem cake extractive (an alcohol extract of neem cake) to be far less toxic to mosquitoes and unstable, with little promise as a mosquito larvicide.

Rao[44] found that neem cake powder, which is used as a fertilizer, has mosquito larvicidal activity under laboratory conditions. To develop a low-cost technology to control mosquito larvae in the paddy ecosystem, he conducted a simulation experiment; 100, 50, 25, and 12.5 g/m² of deoiled neem cake powder were incorporated in mud in simulation trays. Mortality was observed in fourth instar *C. quinquefasciatus* larvae at the 100, 50, and 25 g/m² dosages. At 12.5 g/m², there was no mortality but a high proportion of deformed stages. In trials in paddy fields using 100 kg/ha there was a very low larval density of *Culex* sp. for the first 2 months vs. a high population in the untreated control field. *Anopheles* sp. larvae were controlled to a lesser degree than *Culex* sp.

Kalyamasundaram and Babu[45] tested petroleum ether extracts of a mixture of neem leaves, flowers, seeds, and stems against fourth instar larvae of *C. quinquefasciatus* collected from cess pits (CP), wells (W), and in the field, and then reared in the laboratory (FCRL); 120 mg/l of the extract killed 50% of the CP, 25% of the FCRL, and none of the W larvae. However, the extract had a strong synergistic effect with two insecticides, phenthoate (Cidial®) and fenthion (Lebaycid®), to the dilutions of which the neem extract was added at 1 and 5 mg/l. The following were the LC_{50}s and the synergistic factors (LC_{50} of insecticide: LC_{50} of insecticide with neem) against FCRL larvae: (a) LC_{50} of phenthoate = 3.4 ng/l; (b) LC_{50} of phenthoate with 1 mg/l neem extract = 1.2 ng/l; synergistic factor = 2.8; (c) LC_{50} of phenthoate with 5 mg/l neem = 1.3 ng/l; synergistic factor = 2.6; (d) LC_{50} of fenthion = 1.4 ng/l; (e) LC_{50} of fenthion with 1 mg/l neem = 1.5 ng/l; synergistic factor = 1.35; (f) LC_{50} of fenthion with 5 mg/l neem = 0.5 ng/l; synergistic factor = 2.9.

3. *Anopheles* spp. and Other Mosquitoes

Matemu and Mosha[46] investigated the effect of methanolic and aqueous neem berry extract (NBE) against larvae of *A. gambiae* Giles and *C. quinquefasciatus*. An aqueous dilution of the methanolic extract was somewhat toxic for the larvae, with LC_{50} = 0.02% and LC_{95} = 0.03% for *A. gambiae*, and LC_{50} = 0.05% and LC_{95} = 0.08% for *C. quinquefasciatus*. The methanol extract was slightly better than the aqueous extract against larvae of *C. quinquefasciatus* at concentrations of 0.0125 and 0.025% and about equal to it at 0.05% in the breeding water; at 0.1% the aqueous extract was somewhat more toxic than the methanolic extract. Methanolic extracts from ripe berries gave somewhat better results than those from unripe ones. Adult mosquitoes of both species showed a 24-h mortality of ~80% on topical application with 0.2 µl of the pure methanolic extract per female. When tested at 1.0% in the WHO mosquito kit, NBE was nontoxic for the adults of the two species.

When Gill[18] incubated eggs of *A. stephensi* in a culture medium containing 1 or 2.5% of ANSKE, 1% of the alcoholic extract, or 0.01% azadirachtin, egg hatch was normal but the neonate larvae died within 48 h. Moreover, he found that in third to fourth instar *A. stephensi* larvae 0.01% azadirachtin in the culture medium gave 100% mortality within 24 h, and 0.001% within 7 d, whereas 0.0001% gave incomplete mortality (67%) within 10 d. The last concentration, however, yielded only a few adults (6% of control). When either second to third or fourth instar larvae were introduced into culture media containing crude or sieved aqueous neem suspension, 5 and 1% gave 100% kill of larvae and 0.1% yielded practically no adults.

Some work on mosquitoes was done also with Persian lilac extracts. In larvicide tests conducted by Heal et al.[47] with *A. quadrimaculatus* (Say) and *A. aegypti*, an alcoholic extract and a petroleum ether extract of bark, fruit, leaves, and branches, and a chloroform extract of the aqueous extract of *M. azedarach* were ineffective, as were the petroleum ether extracts of the bark or of the fruits alone.

IV. PARASITES

A. BED BUGS

Chavan[21] tested fraction NP-2 (see above) as a 2% formulation in a Peet-Grady Chamber against bed bugs; 60% mortality was obtained.

B. TRIATOMINE BUGS

Garcia and Rembold[48] investigated the influence of azadirachtin administered in a blood meal on the feeding and ecdysis of fourth instar *Rhodnius prolixus* Stål. nymphs. The antifeedant ED_{50} calculated from the blood ingested was 25 µg/ml and was much higher than the ecydsis inhibiting ED_{50}.* Ecdysone and the JHA, methyl epoxyfarnesoate, antagonized the ecdysis inhibition exerted by azadirachtin, totally and partially, respectively.

Garcia et al.[49] found similar antifeedant ED_{50}s, ranging from 25 to 30 µg/ml blood, for the three compounds azadirachtin A, azadirachtin B, and a synthetic derivative, 7-acetylazadirachtin A, in nymphs of *R. prolixus*. ATP, a well known phagostimulant for hematophagous arthropods, when fed orally in mixture with azadirachtin A, reversed the antifeedant effect of the latter. The effect was especially pronounced with ATP given in excess. It was therefore suggested that azadirachtin blocks the input of phagostimulant receptors. The ecdysis inhibition ED_{50} values were 40, 15, and 450 ng/ml blood for the three compounds, respectively. This shows that the ED_{50} for molt inhibition by the substances was approximately 66 (7-acetylazadirachtin A), 625 (azadirachtin A), and 1735 (azadirachtin B) times lower than the antifeedant ED_{50}s. These data as well as those of Garcia and Rembold support the hypothesis that the feeding inhibition, on the one hand, and molting inhibition, on the other hand, of these compounds are quite distinct from each other as regards their mode of action. The effect of the three com-

---

*   It is not clear from this paper (Garcia and Rembold[48]) how much the ecdysis-inhibiting ED_{50} was, in fact: 0.4 ng/ml (4×10^{-4} µg/ml), as stated in the abstract and on page 940, second paragraph lines 10 to 12; or 40 ng, as can be deduced from Figure 2. Moreover, on page 940, second paragraph, first five lines, it is stated that 5 ng/ml (5×10^{-3} µg/ml) inhibited ecdysis, whereas 2 ng (2×10^{-3} µg/ml) had no effect on this process; so perhaps the ecdysis-inhibiting ED_{50} was 4 ng/ml.

pounds on molting seems to be irreversible, with no molt being observed 5 months (or even for more than 1 year — Garcia et al.[50]) after single treatment and after at least five full refeedings on untreated blood. It is considered that the effect is due to interference of the three azadirachtin derivatives with ecdysone production. Moreover, when Garcia et al.[50] injected 25 ng azadirachtin A into fourth instar nymphs of *R. prolixus* 1, 2, or 3 d after blood feeding, the molting process was interfered with by inhibition of both the onset of epidermal mitosis and the production of ecdysteroids. With doses >50 ng, mortality was high. With doses as low as 2 ng injected, there was 50% molt inhibition, even during an extended larval period of more than 1 year. Since the LD_{50} (within 30 d) was 40 ng, it was 20-fold the EC_{50} for molt inhibition. If, however, the insects were injected later than the first 3 d, e.g., day 7 to day 14 after feeding, that is to say during and after the onset of epidermal mitosis, ecdysis was not affected. It was concluded that a single dose of azadirachtin A injected until day 3 was able to block the onset of epidermal mitosis which is associated with the molting cycle and is triggered by the molting hormone. The ecydsteroid level (determined by radioimmunoassay) in the treated nymphs was too low for induction of ecdysis.

C. FLEAS AND LICE

Rahaman[51] reported that bonnet macaqua (*Macaca radiata*) apes which became infested with fleas (*Pediculus* sp.) and lice (*Ctenocephalaides* sp.) were free of these parasites following application of neem oil.

V. HOUSEHOLD PESTS

A. COCKROACHES

Heal et al.[47] screened the effect of *A. indica* and *M. azedarach* against cockroaches. For the primary screening, a crude aqueous extract was prepared by grinding plant material with a tenfold quantity of distilled water (followed by centrifugation) and using the supernatant liquid. With an *A. indica* extract of branchlets and leaves the following results were obtained: bloodstream injection of the American cockroach, *Periplaneta americana* (L.), caused 50% or higher mortality of both sexes within 3 d. Immersion of the German cockroach, *Blattella germanica* (L.), brought about less than 20% mortality in 4 d. Similarly, injection of the crude aqueous extract of the bark of *M. azedarach* into *P. americana* gave 50% or greater mortality of both sexes within 3 d. The *M. azedarach* aqueous fruit extract injected into *P. americana* brought about 100% paralysis within 2 d, with no subsequent recovery, whereas in an immersion test with *B. germanica*, 20 to 50% mortality was achieved in 4 d. Although an extract of a second sample of the fruit was inactive against both cockroach species, the authors tested *M. azedarach* in a secondary screening which consisted of topical application of extracts to adults of *B. germanica*. An alcoholic extract of bark, fruit, leaves, or branches of *M. azedarach* was effective against the German cockroach (50% mortality at 100 mg/kg). The analogous petroleum ether extract and the chloroform extract of the aqueous extract of bark, fruit, leaves, and branches and the petroleum ether extract of the bark and the fruits alone were inactive against *B. germanica* by topical application.

Fraction NP-2 (see above) was found by Chavan[21] to achieve only 35% mortality against cockroaches (species not given) in a Peet-Grady chamber test.

Further cockroach species were investigated by Adler and Uebel,[52] who tested "Margosan-O", a commercial (Vikwood Ltd., Sheboygan, WI) 40% neem seed extract in 95% ethanol as toxicant, growth inhibitor, or repellent against six species of cockroaches: *Blatta orientalis* L., *Blattella germanica* (L.), *Byrsotria fumigata* Guérin-Méneville, *Grom-*

phadorhina portentosa (Schaum), *Periplaneta americana* (L.), and *Supella longipalpa* (F.). Last instar nymphs of all these species fed Lab-Chow® pellets treated with 0.5 ml Margosan-O per pellet suffered increased mortality and retardation in development. First instar nymphs of *Blatta orientalis, Blatella germanica,* and *S. longipalpa* died after consumption of treated pellets. Last instar *Blatta orientalis* nymphs treated topically with 2 μl of Margosan-O or injected with 0.5 μl showed growth reduction and enhanced mortality, whereas last instar nymphs of *G. portentosa* were not affected at all by 6 μl applied topically. First instar *B. orientalis* nymphs placed on filter paper treated with the neem extract suffered no ill effect. The repellency tests were inconclusive.

The 24-h LD_{50} of azadirachtin injected into the blood stream of last instar male and female nymphs of *P. americana* was found by Quadri and Narsaiah[53] to be 1.5 μg/g body weight. A lower dosage (0.75 μg/g) delayed molting from the 23rd to the 38th d, with no conventional toxic effects, but with molt disturbances (partial molt) leading to death among both sexes. Azadirachtin induced a reduction of 20 to 25% in the hemocyte population 24 h after injection. Disruptive and vacuolated alterations were noticed in both the cytoplasmatic and nuclear regions of the hemocytes. The changes in the composition of the hemolymph were as follows: 25 to 40% reduction in albumin and globulin levels 1 to 24 d after the injection, and 15 to 45% in cholesterol, and 7 to 10% in RNA and DNA levels, 1 to 6 d after injection.

Quadri and Ahmed[54] reported that injection of the LD_{50} of azadirachtin (1.5 μg/g) into male adult *P. americana* resulted in reduction of the volume of the hemolymph which, however, returned to normal values within 16 h. Urea, uric acid, and creatinine content of the hemolymph decreased vs. control during 6 to 8 h after azadirachtin injection. Concurrently, the concentration of these three nitrogenous metabolism waste products increased in the feces excreted by the cockroaches. Naqvi[19] injected a solution containing the neutral fraction of neem winter leaves into the abdomen of *B. germanica* fourth instar nymphs (see also Nizam et al.[55]). Although he claimed that the LD^{50} was 5μg/insect, the result is inconclusive, because the acetone control gave 43% mortality. The author found that inhibition of cholinesterase at the presumed LD_{50} level was 39%, of acid phosphatase, 62%, and of alkaline phosphatase, 74%. Adults with swollen abdomen and decreased wing size were observed.

B. ANTS

Schmidt and Pesel[56] found that the neem extracts AZT-VR-K and MTB/H_2O-K-NR mixed at a rate of 0.25 and 0.5% with a liquid sugar-protein diet had a strong reducing effect on the fecundity of queens of the forest ant, *Formica polyctena* Foerster. Lower concentrations stimulated egg production as did fumigation (filter paper treated with neem extract and fixed to the lid of the receptacles housing the insects). The sterilizing effect by feeding was reversible: when the treatment was stopped, egg laying was resumed.

C. TERMITES

According to Gill[18] the timber of the neem tree is comparatively immune to the attack of termites and other wood-boring insects. He mentions that the protection is popularly attributed to the oil contents of the wood. Also Mangunath[57] states that neem timber is durable; even in exposed situations it is not attacked by white-ants (termites), but he ascribes this to its bitter taste.

Yaga[58,59] tested the antitermitic activity of different woods by offering wood meals to the subterranean termite *Coptotermes formosanus* Shiraki. The relative termite resistance varied in different parts of *M. azedarach.* Although the methanol bark extracts were more effective as regards termite resistance than extracts from leaves, wood, and seeds, the wood was investigated because of its economic importance. The activity was found to be concentrated in the neutral fraction of the methanol extract. Two termicidal compounds were isolated in the crystalline state. One was the limonoid nimbolin A, whereas the exact structure of the other, $C_{23}H_{38}O_5$, remained unknown.

Butterworth and Morgan[60] cited work done at the Forest Products Research Laboratory at Princes Risborough, England, according to which impregnation of Ilomba wood (*Pycnanthus angolensis*) with azadirachtin had no effect on the feeding or the survival of the Mediterranean moist wood termite (*Reticulitermes santonensis* A. J. Flux). Jacobson[24] reported (personal communications from F. L. Carter) that in tests carried out at the USDA, a 3% neem seed emulsion repelled adults of the termite *R. virginicus* Banks.

Dutta[61] mixed soil in pot trials with different rates of neem cake powder (60, 48, 36, 24, and 12 kg/ha). Termites (*Microtermes* sp.) exposed to the treated soils died within 24 h even at the lowest concentration, which retained its residual toxicity for 4 weeks.

D. CRICKETS

Jacobson[24] reported that in USDA tests neem repelled adults of the house cricket, *Acheta domesticus* (L.). Warthen and Uebel[62] at the USDA found that when newly hatched house crickets were fed a diet containing 1, 10, or 25 mg/kg azadirachtin, they gained less weight and were retarded in development as compared with the controls. The average weight of the crickets was inversely proportional to the concentration of azadirachtin in the diet. In seventh instar female nymphs fed higher azadirachtin concentrations (10, 50, or 100 mg/kg), too, the amount of feeding and gain in weight was inversely proportional to the azadirachtin concentration in the diet. In both cases, fewer insects survived on the azadirachtin diets.

When first instar nymphs were fed for 6 weeks on a diet containing 10 mg/kg azadirachtin and then transferred and kept for a further 6 weeks on an untreated diet, they recovered quickly from the strong antifeedant effect (average weight of crickets fed azadirachtin for 6 weeks — 10 mg; of control animals — 67 mg): the mean weight of the crickets transferred from the azadirachtin to the untreated diet increased rapidly and by the 12th week even surpassed that of the control group. Also in this experiment crickets fed on an azadirachtin diet gained weight and reached adulthood more slowly than crickets on an untreated diet.

Excision of the maxillary and labial palpi of sixth and seventh instar nymphs did not reduce the antifeedant effect of 50 mg/kg azadirachtin, which means that receptors other than, or in addition to, those present on the palpi, must be responsible for the antifeedant effect of azadirachtin; similarly operated upon crickets kept on untreated diet gained weight normally.

First instar crickets kept on filter paper treated with various dosages of azadirachtin ranging from 1 to 25 $\mu g/cm^2$ showed feeding and growth inhibition during the first 6 weeks; later the response became erratic. Topical application of 50 μg azadirachtin per seventh instar nymph resulted in emergence of only 20% adults of the treated males and 17% of the treated females vs. 88 and 83%, respectively, in the acetone-only treated controls. Similar results were obtained with the eighth instar nymphs topically treated with 25 μg azadirachtin/nymph (azadirachtin treatment male yield — 14%, female yield — 22%, both male and female controls — 100%).

Incidentally, in order to achieve complete protection of 2 × 2 cm^2 corn leaf squares against females of field crickets (*Gryllus pennsylvanicus* Burmeister), Adler and Uebel[63] had to apply 20 μl/square of the commercial neem extract Margosan-O at the high concentration of 10%.

E. FABRIC-DAMAGING INSECTS

Mangunath[57] states that neem leaves are bitter and have a faint but characteristic unpleasant smell. Dried in the shade, they are commonly placed in India in books, papers, and clothes to protect them from moths, etc. Jacobson[24] reported (personal communication from R. E. Bry) that cloth samples impregnated with the rather high concentration of 0.5% azadirachtin were satisfactorily protected from larvae of the clothes moth, *Tineola bisselliella* (Hummel).

Heal et al.[47] tested extracts of *M. azedarach* against larvae of the clothes moth and of the black carpet beetle *Attagenus piceus* (Oliv.), by a fabric protection test. An alcoholic extract prepared of bark, fruit, leaves, and branches was active against both species (less than 10 mg of weight loss with 2% deposits on wool). The analogous petroleum ether extract of bark, fruit, leaves, and branches and of the bark or the fruits alone, and the chloroform extract of the aqueous extract, were active only against the carpet beetle.

Perte[64] observed that leaves and extracts of *Azadirachta indica* were of some value in preventing damage by the furniture carpet beetle *Anthrenus vorax* Waterh. Jacobson[24] mentioned (personal communication from R. E. Bry) black carpet beetle, *Attagenus megatoma* F. = [*A. multicolor* (Brahm)] = *A. piceus*, adults as being repelled by a 3% neem emulsion in USDA tests.

VI. BENEFICIAL INSECTS

A. BEES

Rembold et al.[65] treated third instar worker larvae of the honey bee, *Apis mellifera* L., by topical application with different concentrations of NN 18-704, a neem fraction purified on a silica gel column. The stock solution was equivalent to 1% neem seed kernel in methanol (w/v). This solution was tested undiluted or diluted with methanol at 1:1, 1:5, 1:10, and 1:50. In all these treatments the larval weight gain was no different from the untreated control, which indicated that fraction NN 18-704 had no antifeedant effect. Pupal development was normal except at the 1:1 ratio, in which the survival of the larvae to the pupal and adult stages was 59 and 45%, respectively, of that of the untreated control. Larval abnormalities were observed which caused the death of the larvae affected. Caste determination (percentage of intercastes and queens among the total number of adults) was lower at the 1:1 ratio (44%) than in the untreated control (70%).

Comparison of the effect of the neem extract on honey bee larvae and larvae of *Epilachna varivestis* Muls. and *Ephestia kuehniella* Zell. showed that one tenth of the dose effective for the two latter insect species caused the metamorphic disturbances in *A. mellifera*.

Sharma et al.[66] in Rembold's laboratory assayed four further neem seed fractions purified by column chromatography and preparative TLC (NN 18-701, NN 18-705, NN 18-79a, and NN 18-79b) against third instar larvae of the honey bee, as above. Topical application of the fractions dissolved in methanol — 30% aliquots of the first two fractions and 10% of the last two, corresponding to neem seed — was conducted. Adult emergence (corrected for control) after treatment with NN 18-705 and NN 18-79a was 57 and 49%, respectively. Both fractions were also partial inhibitors of larval feeding. On testing the four fractions after a further dilution (1:1 with methanol), it was found that adult emergence (in all cases) and larval weight gain (except with NN 18-701) were dose-dependent. When the percentage survival at different stages from fifth instar larvae to the pupal stage was recorded, maximum mortality occurred in the 2-d interval between prepupa and pupa in all the fractions except NN 18-79b. During this period the mortality was highest immediately before larval-pupal ecdysis, which indicates disturbances in metamorphosis. No such mortality was observed with NN 18-701, the least effective of the fractions. Similar results were found by Rembold et al.,[67,68] when treating third instar *A. mellifera* larvae with azadirachtin itself by topical application. Survival to the adult stage was greatly reduced by doses of 0.25 and 0.5 μg/larva azadirachtin (58 and 22% corrected survival, respectively). There was no feeding inhibition due to doses up to 0.25 μg and only a very slight reduction in weight gain vs. the control at 0.5 μg/larva azadirachtin, but at these two dosages abnormal larvae were observed 24 to 48 h after application.

In view of the results of Rembold's group it was of interest to elucidate whether NSKE caused any damage to bee colonies. Schmutterer et al.[69] therefore tested a 25% emulsion concentrate of the extract AZT-VR-K at a dilution of 2 ml/l water against honey bees in a field cage experiment and in a regular field trial. Pollen- and nectar-producing plants, wild mustard (*Sinapis alba*), tansy phacelia (*Phacelia tanacetifolia*), and summer rape (*Brassica napus*), grown in 4-m² plots in field cages, were repeatedly (three sprays at 3-d intervals) sprayed at a rate of 60 ml/m², and bee behavior and colony development were observed. Only in very small bee colonies consisting of a queen and 200 to 300 workers was some damage observed in the population, mainly in those exposed to treated phacelia which blossomed at the time of the experiment. A number of young bees (~50) were unable to hatch from the cells after biting off the lids. Furthermore, 10 d after the phacelia was sprayed, a few young bees had crippled wings or parts of the pupal exuvium adhering to the abdomen.

The bees that were unable to emerge and died in the cells were removed with their cells by the workers. Later on, the damaged bees disappeared. With another small population, with about 200 workers more than the first one, only about 20 bees were damaged. The queens of both small populations continued normal oviposition during the experiments. No negative effects on the ovarioles of the queens were found in a histopathological examination. In a bigger but still small population — about 3000 workers — no damage was found, nor was there any damage to the brood in the field trial sprayed once with 70 ml/m². There was no repellency effect for the bees with the treated phacelia and foraging bees did not evince any damage or unusual behavior; this was true also with bumble bees. The authors believe that under field conditions the possibility of serious damage to bees by neem products is not expected. However, since neem may not be completely safe to bees, the authors recommended that high concentrations of neem extract should not be sprayed on flowers which are visited by numerous honey bees. Jacobson et al.[70] cited results which show that high concentrations (1% w/v) of a crude neem seed ethanol extract in 50% (w/v) sugar syrup cause a 50% reduction of feeding by honey bees in laboratory cages. Spatial tests indicated repellency at 1 mg/cm². It is concluded that the extract functions as a gustatory but not as an olfactory repellent. However, olfactometer tests conducted by Gupta[71] show *Azadirachta indica* essential oil to be 63 to 80% repellent against workers of the little honey bee, *Apis florea* F., in the concentration range between 0.0625 and 0.5%.

B. SILKWORM

Koul et al.[72] injected azadirachtin into fifth instar larvae of *Bombyx mori* (L.) from day 0 until day 6 after molt. The effect obtained depended on the age of the larvae: whereas defective pupae developed from day 0- to day 3-treated larvae, larvae injected on days 4 to 6 failed to pupate, had epidermal black bands around the segments, and died in the larval state on day 12. It was shown in *in vitro* and by ligation experiments that azadirachtin had no direct effect on prothoracicotropic hormone and prothoracic gland secretion. The authors suggest that azadirachtin has more than one site of action. They believe that the azadirachtin-produced inhibitions are due to a general blockage of factors in the central nervous system and that the effect depends on the balance of developmental hormones, e.g., prothoracicotropic hormone, at the time of treatment.

VII. CONCLUSIONS

Much less work has been done with neem products on insects of medical and hygienic importance, household pests, and beneficial insects associated with man, than on agricultural and stored-product pests. Very often concentrations needed to control insects affecting man and animal with neem are rather high (see, for example, mosquitoes). Therefore it is hard to envision widespread use of neem products against this sector of insects. However, the physiological studies conducted, especially with the honey bee, *Rhodnius prolixus* and *Calliphora vicina*, have yielded findings highly useful in elucidating the mode of action of neem in general and of azadirachtin in particular in the class Insecta.

REFERENCES

1. **Howaldt, T.,** *Azadirachta indica* A. Juss; *Tamarindus indica* L., Occurrence, Forest Cultivation and Possible Uses, Ph.D. thesis, University of Hamburg, West Germany, 1980 (in German).
2. **Anon.,** *Azadirachta indica,* in *Firewood Crops: Shrubs and Tree Species for Energy Production,* National Academy of Sciences, Washington, D.C., 1980, 114.
3. **Butterworth, J. H. and Morgan, E. D.,** Isolation of a substance that suppresses feeding in locusts, *J. Chem. Soc., Chem. Commun.,* 23, 1968.
4. **Lavie, D., Jain, M. K., and Shpan-Gabrielith, S. R.,** A locust phagorepellent from two *Melia* species, *J. Chem. Soc., Chem. Commun.,* 910, 1967.
5. **Warthen, J. D., Jr., Uebel, E. C., Dutky, S. R., Lusby, W. R., and Finegold, H.,** Adult house fly feeding deterrent from neem seeds, U.S. Department of Agriculture, Science and Education Administration, Agric. Res. Results, Ser. ARR-NE-2, 1978.
6. **Forster, H.,** Isolation of Azadirachtins from Neem (*Azadirachta indica*) and Radioactive Labelling with Azadirachtin A, Ph.D. thesis, University of Munich, West Germany, 1983 (in German).
7. **Rembold, H., Forster, H., Czoppelt, Ch., Rao, P. J., and Sieber, K.-P.,** The azadirachtins, a group of insect growth regulators from the neem tree, in *Natural Pesticides from the Neem Tree and Other Tropical Plants,* Schmutterer, H. and Ascher, K. R. S., Eds.,

GTZ Press, Eschborn, West Germany, 1984, 153.
8. **Morgan, E. D. and Thornton, M. D.,** Azadirachtin in the fruit of *Melia azedarach,* *Phytochemistry,* 12, 391, 1973.
9. **Zanno, P. R., Miura, I., Nakanishi, K., and Elder, D. L.,** Structure of the insect phagorepellent azadirachtin. Application of PRFT/CWD carbon-13 nuclear magnetic resonance, *J. Am. Chem. Soc.,* 97, 1975, 1975.
10. **Bilton, J. N., Broughton, H. B., Ley, S. V., Lidert, Z., Morgan, E. D., Rzepa, H. S., and Sheppard, R. N.,** Structural reappraisal of the limonoid insect antifeedant azadirachtin, *J. Chem. Soc., Chem. Commun.,* 968, 1985.
11. **Kraus, W., Bokel, M., Klenk, A., and Pöhnl, H.,** The structure of azadirachtin and 22,23-dihydro-23β-methoxyazadirachtin, *Tetrahedron Lett.,* 26, 6435, 1985.
12. **Ketkar, C. M.,** Final Technical Report — Utilization of Neem (*Azadirachta indica* A. Juss) and Its By-products, Bombay, 1976.
13. **Warthen, J. D., Jr.,** *Azadirachta indica:* a source of insect feeding inhibitors and growth regulators, U.S. Department of Agriculture, Science and Education Administration, Agric. Rev. Manuals, Ser. ARM-NE-4, 1979.
14. **Jain, H. K.,** *Neem in Agriculture,* Indian Agricultural Research Institute, New Delhi, 1983.
15. **Jacobson, M.,** The neem tree: Natural resistance par excellence, in *Natural Resistance of Plants to Pests — Roles of Allelochemicals,* Green, M. B. and Hedin, P. A., Eds., ACS Symp. Ser. No. 296, Amer-

ican Chemical Society, Washington, D.C., 220, 1987.

16. **Quadri, S. S. H. and Rao, Bh. B.,** Effect of combining some indigenous plant seed extracts against household insects, *Pesticides (Bombay),* 11(12), 21, 1977.

17. **Quadri, S. S. H. and Rao, Bh. B.,** Effect of oleoresin in combination with neem seed and garlic clove extracts against household and stored products pests, *Pesticides (Bombay),* 14(3), 11, 1980.

18. **Gill, J. S.,** Studies on Insect Feeding Deterrents with Special Reference to the Fruit Extracts of the Neem Tree, *Azadirachta indica* A. Juss, Ph.D. thesis, University of London, England, 1972.

19. **Naqvi, S. N. H.,** Biological evaluation of fresh neem extracts and some components with reference to abnormalities and esterase activity in insects, *Proc. 3rd Int. Neem Conf.,* Nairobi, Kenya, 1986, Schmutterer, H. and Ascher, K. R. S., Eds., GTZ Press, Eschborn, West Germany, 1987, 315.

20. **Saeed, S. A. and Naqvi, S. N. H.,** Toxicity of NfC (neem extract) against *Musca domestica* and their [its] effect on esterase activity, *Pak. J. Entomol., Karachi,* 1, in press; cited in **Naqvi, S. N. H.,** *Proc. 3rd Int. Neem Conf.,* Nairobi, Kenya, 1986, Schmutterer, H. and Ascher, K. R. S., Eds., GTZ Press, Eschborn, West Germany, 1987, 315.

21. **Chavan, S. R.,** Chemistry of alkanes separated from leaves of *Azadirachta indica* and their larvicidal/insecticidal activity against mosquitoes, in *Natural Pesticides from the Neem Tree and Other Tropical Plants,* Schmutterer, H. and Ascher, K. R. S., Eds., GTZ Press, Eschborn, West Germany, 1984, 59.

22. **Ahmed, S. M., Chander, H., and Pereira, J.,** Insecticidal potential and biological activity of Indian indigenous plants against *Musca domestica* L., *Int. Pest Control,* 23, 170, 1981.

23. **Perera, H. C.,** Indigenous drugs in the treatment of livestock, *Indian Vet. J.,* 17, 261, 1941.

24. **Jacobson, M.,** Neem research in the U.S. Department of Agriculture: chemical, biological and cultural aspects, in *Natural Pesticides from the Neem Tree (Azadirachta indica A. Juss),* Schmutterer, H., Ascher, K. R. S., and Rembold, H., Eds., GTZ Press, Eschborn, West Germany, 1981, 33.

25. **Gaaboub, I. A. and Hayes, D. K.,** Biological activity of azadirachtin, component of the neem tree, inhibiting molting in the face fly, *Musca autumnalis* De Geer (Diptera: Muscidae), *Environ. Entomol.,* 13, 803, 1984. 1984.

26. **Gaaboub, I. A. and Hayes, D. K.,** Effect of larval treatment with azadirachtin, a molting inhibitory component of the neem tree, on reproductive capacity of the face fly, *Musca autumnalis* De Geer (Diptera:Muscidae), *Environ. Entomol.,* 13, 1639, 1984.

27. **Hobson, R. P.,** Sheep blow-fly investigations. VIII. Observations on larvicides and repellents for protecting sheep from attack, *Ann. Appl. Biol.,* 27, 527, 1940.

28. **Rice, M. J., Sexton, S., and Esmail, A. M.,** Antifeedant phytochemical blocks oviposition by sheep blowfly, *J. Aust. Entomol. Soc.,* 24, 16, 1985.

29. **Bidmon, H.-J., Käuser, G., Möbus, P., and Koolman, J.,** Action of azadirachtin on blowfly larvae and pupae, in *Proc. 3rd Int. Neem Conf.,* Nairobi, Kenya, 1986, Schmutterer, H. and Ascher, K. R. S., Eds., GTZ Press, Eschborn, West Germany, 1987, 253.

30. **Radwanski, S.,** Neem tree. III. Further uses and potential uses, *World Crops Livestock,* 29, 167, 1977.

31. **Schmutterer, H. and Zebitz, C. P. W.,** Effect of methanolic extracts from seeds of single neem trees of African and Asian origin, on *Epilachna varivestis* and *Aedes aegypti,* in *Natural Pesticides from the Neem Tree and Other Tropical Plants,* Schmutterer, H. and Ascher, K. R. S., Eds., GTZ Press, Eschborn, West Germany, 1984, 83.

32. **Ermel, K., Pahlich, E., and Schmutterer, H.,** Comparison of the azadirachtin content of neem seeds from ecotypes of Asian and African origin, in *Natural Pesticides from the Neem Tree and Other Tropical Plants,* Schmutterer, H. and Ascher, K. R. S., Eds., GTZ Press, Eschborn, West Germany, 1984, 91.

33. **Zebitz, C. P. W.,** Effect of some crude and azadirachtin-enriched neem (*Azadirachta indica*) seed kernel extracts on larvae of *Aedes aegypti, Entomol. Exp. Appl.,* 35, 11, 1984.

34. **Zebitz, C. P. W.,** Effects of neem (*Antelaea azadirachta*)-products on mosquitoes, *Verhandl. Dtsch. Zool. Ges.,* 78, 199, 1985 (in German).

35. **Zebitz, C. P. W.,** Potential of neem seed kernel extracts in mosquito control, in *Proc. 3rd Int. Neem Conf.,* Nairobi, Kenya, 1986, Schmutterer, H. and Ascher, K. R. S., Eds., GTZ Press, Eschborn, West Germany, 1987, 555.

36. **Hellpap, C. and Zebitz, C. P. W.,** Combined application of neem seed extract with *Bacillus thuringiensis* to control *Spedoptera frugiperda* and *Aedes togoi, Z. Angew. Entomol.,* 101, 515, 1986 (in German).

37. **Naqvi, S. N. H. and Ahmed, S. O.,** Toxicity and effect of some neem products on *Aedes aegypti* L. (PCSIR) strain, *Pak. J. Entomol., Karachi,* in press; cited by **Navqi, S. N. H.,** *Proc. 3rd Int. Neem Conf.,* Nairobi, Kenya,

1986, Schmutterer, H. and Ascher, K. R. S., Eds., GTZ Press, Eschborn, West Germany, 1987, 315.

38. **Mwangi, R. W. and Rembold, H.,** Growth-regulating activity of *Melia volkensii* extracts against the larvae of *Aedes aegypti*, in *Proc. 3rd Int. Neem Conf.*, Nairobi, Kenya, 1986, Schmutterer, H. and Ascher, K. R. S., Eds., GTZ Press, Eschborn, West Germany, 1987, 669.

39. **Singh, R. P.,** Effect of water extract of deoiled neem kernel on second instar larvae of *Culex fatigans* Wiedemann, *Neem Newslett.*, 1, 16, 1984.

40. **Chavan, S. R., Deshmukh, P. B., and Renapurkar, D. M.,** Investigations of indigenous plants for larvicidal activity, *Bull. Haffkine Inst.*, 7, 23, 1979.

41. **Attri, B. S. and Prasad, R.,** Neem oil extractive, an effective mosquito larvicide, *Indian J. Entomol.*, 42, 371, 1980.

42. **Jotwani, M. G. and Srivastava, K. P.,** A review of neem research in India in relation to insects, in *Natural Pesticides from the Neem Tree and Other Tropical Plants*, Schmutterer, H. and Ascher, K. R. S., Eds., GTZ Press, Eschborn, West Germany, 1984, 43.

43. **Attri, B. S. and Prasad, R.,** Studies on the pesticidal values of neem oil by-products, *Pestology*, 4(3), 16, 1980.

44. **Rao, D. R.,** Assessment of neem cake powder, *Azadirachta indica* A. Juss (Meliaceae), as a mosquito larvicide, Abstr. Symp. on Alternatives to Synthetic Insecticides in Integrated Pest Management Systems, New Delhi, 1987, 23.

45. **Kalyanasundaram, M. and Babu, C. J.,** Biologically active plant extracts as mosquito larvicides, *Indian J. Med. Res. (Suppl.)*, 76, 102, 1982.

46. **Matemu, D. P. and Mosha, F. W.,** Toxic effects of neem (*Azadirachta indica*) berry extract on mosquitoes, *Neem Newslett.*, 3, 44, 1986.

47. **Heal, R. E., Rogers, E. F., Wallace, R. T., and Starnes, O.,** A survey of plants for insecticidal activity, *Lloydia*, 13, 89, 1950.

48. **Garcia, E. S. and Rembold, H.,** Effects of azadirachtin on ecdysis of *Rhodnius prolixus*, *J. Insect Physiol.*, 30, 939, 1984.

49. **Garcia, E. S., Azambuya, de P., Forster, M., and Rembold, H.,** Feeding and molt inhibition by azadirachtins A, B, and 7-acetyl-azadirachtin A in *Rhodnius prolixus* nymphs, *Z. Naturforsch.*, 39c, 1155, 1984.

50. **Garcia, E. S., Uhl, M., and Rembold, H.,** Azadirachtin, a chemical probe for the study of moulting processes in *Rhodnius prolixus*, *Z. Naturforsch.*, 41c, 771, 1986.

51. **Rahaman, H.,** Ectoparasitism in *Macaca radiata*, *Lab. Anim. Sci.*, 25, 505, 1975.

52. **Adler, V. E. and Uebel, E. C.,** Effects of a formulation of neem extract on six species of cockroaches (Orthoptera:Blaberidae, Blattidae and Blattellidae), *Phytoparasitica*, 13, 3, 1985.

53. **Quadri, S. S. H. and Narsaiah, J.,** Effect of azadirachtin on the molting processes of last instar nymphs of *Periphaneta americana* (Linn.), *Indian J. Exp. Biol.*, 16, 1141, 1978.

54. **Quadri, S. S. H. and Ahmed, M.,** Effect of indigenous plant pesticides on the urea, uric acid and creatinine contents of fecal matter and haemolymph of the cockroach *Periplaneta americana* (Linn.), *Indian J. Exp. Biol.*, 17, 95, 1979.

55. **Nizam, S., Naqvi, S. N. H., and Ahmed, I.,** Toxicity of reserpine and NfB (neem extract) against *Blattella germanica* L. and their effect on esterases, *J. Sci. Karachi Univ.*, in press; cited by **Naqvi, S. N. H.,** *Proc. 3rd Int. Neem Conf.*, Nairobi, Kenya, 1986, Schmutterer, H. and Ascher, K. R. S., Eds., GTZ Press, Eschborn, West Germany, 1987, 315.

56. **Schmidt, G. H. and Pesel, E.,** Studies of the sterilizing effect of neem extracts in ants, in *Proc. 3rd Int. Neem Conf.*, Nairobi, Kenya, 1986, Schmutterer, H. and Ascher, K. R. S., Eds., GTZ Press, Eschborn, West Germany, 1987, 361.

57. **Mangunath, B. L., Ed.,** *The Wealth of India — Raw Materials*, Vol. 1, The Council of Scientific and Industrial Research, New Delhi, 1948, 140.

58. **Yaga, S.,** On the termite-resistance of Okinawan timbers, *Ryukyu Daigaku Nogakubu Gakujutsu Hokoku (Bull. Coll. Agric. Ryukyus University)*, 25, 555, 1978.

59. **Yaga, S.,** On the termite resistance of Okinawan timbers. VI. Termicidal substances from *Melia azedarach* L. *Ryukyu Daigaku Nogakubu Gakujutsu Hokoku (Bull. Coll. Agric., Ryukyus University)*, 26, 494, 1980.

60. **Butterworth, J. H. and Morgan, E. D.,** Investigation of the locust feeding inhibition of the seeds of the neem tree, *Azadirachta indica, J. Insect Physiol.*, 17, 969, 1971.

61. **Dutta, N.,** cited in **Ketkar, C. M.,** Final Technical Report — Utilization of Neem (*Azadirachta indica* A. Juss) and Its By-products, Bombay, 1976.

62. **Warthen, J. D., Jr. and Uebel, E. C.,** Effect of azadirachtin on house crickets, *Acheta domesticus*, in *Natural Pesticides from the Neem Tree and Other Tropical Plants*, Schmutterer, H. and Ascher, K. R. S., Eds., GTZ Press, Eschborn, West Germany, 1981, 137.

63. **Adler, V. E. and Uebel, E. C.,** Antifeedant bioassays of neem extract against the Carolina grasshopper, walkingstick, and field cricket, *J. Environ. Sci. Health*, 19A, 393, 1984.

64. **Perte, S. L.,** Short notes and exhibits, *Indian J. Entomol.,* 4, 94, 1942; cited by **Ketkar, C. M.,** Final Technical Report — Utilization of Neem (*Azadirachta indica* A. Juss) and Its By-products, Bombay, 1976.

65. **Rembold, H., Sharma, G. K., Czoppelt, Ch., and Schmutterer, H.,** Evidence of growth disruption in insects without feeding inhibition by neem seed fractions, *Z. Pflanzenkr. Pflanzenschutz,* 87, 290, 1980.

66. **Sharma, G. K., Czoppelt, Ch., and Rembold, H.,** Further evidence of insect growth disruption by neem seed fractions, *Z. Angew. Entomol.,* 90, 439, 1980.

67. **Rembold, H., Sharma, G. K., and Czoppelt, Ch.,** Growth-regulating activity of azadirachtin in two holometabolous insects, in *Natural Pesticides from the Neem Tree (Azadirachta indica A. Juss),* Schmutterer, H., Ascher, K. R. S., and Rembold, H., Eds., GTZ Press, Eschborn, West Germany, 1981, 121.

68. **Rembold, H., Sharma, G. K., Czoppelt, Ch., and Schmutterer, H.,** Azadirachtin: a potent insect growth regulator of plant origin, *Z. Angew. Entomol.,* 93, 12, 1982.

69. **Schmutterer, H. and Holst, H.,** Investigations of the effect of enriched and formulated neem extracts AZT-VR-K on the honeybee *Apis mellifera* L., *Z. Angew. Entomol.,* 103, 208, 1987 (in German).

70. **Jacobson, M., Stokes, J. B., Warthen, J. D., Jr., Redfern, R. E., Reed, D. K., Webb, R. E., and Telek, L.,** Neem research in the U.S. Department of Agriculture: an update, in *Natural Pesticides from the Neem Tree and Other Tropical Plants,* Schmutterer, H. and Ascher, K. R. S., Eds., GTZ Press, Eschborn, West Germany, 1984, 31.

71. **Gupta, M.,** Essential oils: a new source of bee repellents, *Chem. Ind. (London),* (5), 162, 1987.

72. **Koul, O., Amanai, K., and Ohtaki, T.,** Effect of azadirachtin on the endocrine events of *Bombyx mori, J. Insect Physiol.,* 33, 103, 1987.

8 Pharmacology and Toxicology of Neem

Martin Jacobson
U.S. Department of Agriculture (Retired)
Silver Spring, Maryland

TABLE OF CONTENTS

I. INTRODUCTION

The neem tree, *Azadirachta indica* A. Juss (synonym *Melia azadirachta* L.), also commonly known as "nim" or "margosa" has sometimes been confused with the chinaberry tree, *Melia azedarach* L., in the literature. Whereas the latter, also commonly known as "China tree," "Chinese umbrella tree," and "pride of India," grows wild and is cultivated for shade or ornament worldwide, the neem tree in the wild is fairly well restricted to the dry and hot climates of Asia and Africa, although it is presently being cultivated in selected areas of the Western Hemisphere.[1,2] Although the various parts of neem and the fruits of chinaberry have been used as insecticides and parasiticides and are referred to in Ayurvedic medicine,[2-5] chinaberry is no longer used for these purposes because it has been shown to be highly toxic to warm-blooded animals.[5-9]

In sharp contrast with the decline of the chinaberry as a source of pesticides and medicinals, the continued use of neem for these purposes has shown phenomenal ascendancy, especially within the past 10 years.[10-15]

The medicinal properties ascribed to the neem tree — especially to the fruit, bark, and leaves — are legion since ancient times. They are mentioned in the earliest Sanskrit medical writings.[16-19] Excellent reviews of the older literature are those by Radwanski,[3] Mitra,[20,21] Das,[22] and Jotwani and Srivastava.[23] Medicinal preparations from various parts of the tree, as well as its seed oil, attracted the attention of European physicians practicing in India in the 19th century, who then recorded the properties of neem in Government records.[20,24] For many years the people of India used neem twigs as toothbrushes, and toothpastes containing neem preparations are available there now on the open market.[25,26] Neem is utilized in the manufacture of bath soaps to safeguard the skin from microbial infection, and these products are likewise available today in a number of Asian countries.[27,28]

According to Ahmed,[27] "neem leaf juice and decoction have antihelminthic, antiseptic, diuretic, emmenagogic, emollient, and purgative properties and are also used to treat eczema and ulcers. The plant is also used to treat blood disorders, hepatitis, and eye diseases, and it is thought to possess antisyphilitic properties." Lal and Yadov[29] report that local applications of a paste prepared from tender neem leaves help to remove facial pimples, and a mixture prepared by grinding 50 g of wet neem seeds with 150 g of "misri" (crystalline sugar) made into tablets of 150 mg each is used to treat boils and pimples when three tablets are taken three times a day.

Hartwell,[30] in his encyclopedic volume on plants used to treat cancer, cites a number of published sources advocating the use of neem preparations. Poultices of neem leaves were applied to glandular tumors[31] and abdominal tumors in India.[32,33] Neem oil is also reported to have anti-inflammatory properties useful in the treatment of ulcers.[20,34,35]

Patrao[36] cites the use of aqueous extracts of neem leaves for treating constipation, diabetes, indigestion, itch, pyorrhoea, sleeplessness, and stomachache. Duke and Wayne[37] report various folk remedies for the use of neem formulations as alteratives, antiseptics, astringents, emollients, febrifuges, anodynes, diuretics, parasiticides, pediculicides, purgatives, sedatives, stomachics, and tonics, as well as treatments for boils, burns, cholera, fever, gingivitis, heat rash, malaria, measles, nausea, rheumatism, snakebite, and syphilis!

Given the great mass of laudatory reports on the medicinal uses of neem, it was inevitable that physicians and other scientists should conduct research directed toward the determination of the truth or falsity of the foregoing claims. Amazingly, the following discussions of such pharmacological research serve to substantiate many of these claims.

II. PHARMACOLOGY

A. DERMATOLOGICAL EFFECTS

The Pharmacology Department of King George's Medical College in Lucknow, India successfully used a lotion made from the extract of dried neem leaves to cure eczema, ringworm, and scabies in clinical evaluation tests.[38] Acute cases were cured permanently in 3 to 4 d by local application of the lotion; chronic cases required about a fortnight. Administration of the extract in capsule form to patients with ringworm was also highly effective using a single dose of six capsules, but the dose/capsule was not specified.

Edema induced in rat paws by carrageenin (a marine colloid) injection was considerably reduced by oral administration of 2 ml of a solution of leaf methanolic extract in 2% sodium carbonate, each animal receiving 400 or 800 mg/kg body weight.[39,40]

B. ANTIPYRETIC EFFECTS

Male rabbits were made hyperpyretic by intravenous injection of 1 μg of pyrogenic lipopolysaccharide. After 1 h several of the animals received an oral dose of 400 mg of neem leaf methanolic extract in 5 ml of 2% sodium carbonate/kg body weight. An equal number of control animals received 200 mg/kg of acetylsalicylic acid. Rectal temperatures of all rabbits were recorded before administration and every 30 min thereafter. After 3 h fever induced by the pyrogen was reduced by 25% using acetylsalicylic acid and by 15.7% using neem extract.[39]

C. DIURETIC EFFECTS

A saline solution of the sodium salt of nimbidinic acid, a component of neem seeds,[35] was administered to several adult dogs by intravenous injection (200 mg/kg) or by stomach tube. Urine samples following administration showed no evidence of protein or reducing substances, but the amount of sodium chloride and potassium was increased. No increase in uric acid excretion was detected and the pH remained unchanged.[41] Intravenous injection of a saline or glucose solution (1 mg/kg) prior to the administration of sodium nimbidinate provoked a prompt, long-lasting diuresis. The sodium salt at 2 to 20 mg was a far more potent diuretic than urea at 12 g. The substance appears to be excreted in the urine.

D. CIRCULATORY EFFECTS

Perfusion of the heart of a frog with Ringer's solution containing 10^{-5} g of sodium nimbidinate/ml for 45 min did not affect the rate, rhythm, or force of contraction.[42] Concentrations of 10^{-4} or 10^{-3} g/ml had no effect on the heart, but allowing a solution of 10^{-3} g/ml to act for 1 min diminished the force of contraction and produced a change in rhythm. Perfusion of an isolated rabbit heart with Tyrode's solution containing 10^{-4} g/ml caused no reaction, but concentrations of 10^{-3} or 10^{-2} g/ml caused a progressive diminution in the force of contraction and ultimate stoppage in diastole. Intravenous injection of 0.1 to 1.0 mg sodium nimbidinate/kg into anesthetized rats had no effect on blood pressure or respiration, but 10 mg/kg produced a perceptible fall in pressure within 10 s, with proportionately greater reduction using 20 and 40 mg/kg. The reduction in blood pressure was not antagonized by atropine or promethazone.

Mitra[20] found that neem seed oil could be used for relieving tissue edema in congestive heart failure. Pillai and Santhakumari[43] reported that feeding the oil (2.5 ml/kg) or nimbidin (200 mg/kg), one of its bitter principles, to fasting rabbits caused significant hypoglycemic activity through reduction of the blood sugar level by 24 and 26%, respectively, within 5 h. Nimbidin was fed as a 10% ethanol solution. An aqueous extract of fresh neem leaves containing 1 g of leaves/ml of decoction had no effect at 1, 5, or 10 ml/

kg. In glucose tolerance tests, 1 ml of leaf decoction, 2.5 ml of seed oil, and 200 mg of nimbidin each significantly delayed peak rise in blood sugar after glucose administration. Both the oil and nimbidin were half as potent as tolbutamide.

Nimbolide and 3-deacetylsalannin, tetranortriterpenoids isolated from the chloroform-soluble portion of an alcoholic extract of neem leaves, have shown hypotensive activity.[44]

A crude aqueous extract of neem leaves was studied for its effects on the cardio-vascular system of anesthetized guinea pigs and rabbits.[45] Intravenous injection of 5, 10, 20, 40, 80, 100, and 200 mg/kg, administered at 30-min intervals, induced profound hypotension and a minimal negative chronotropic effect, which increased with dose. In one rabbit, 200 mg reduced the heart rate from 280 to 150 beats/min. The extract also exhibited a weak antiarrhythmic activity in rabbits against ouabain-induced dysrhythmia.

Nine adult dogs were subjected to partial pancreatectomy and a corresponding number were subjected to complete pancreatectomy.[46] Several dogs of each group were fed 8 to 10 drops of neem seed oil in gelatin capsules daily. The average survival times were 70 and 15 d, respectively, for the partially and completely pancreatectomized dogs; control dogs (no neem oil) survived for 40 d. Skin wounds, hair loss, rough skin coat, and clinical signs of dehydration were seen in partially pancreatectomized animals. Absence of pancreatic function (as demonstrated by trypsin test) was seen in the completely pancreatectomized animals, but some function was shown by the partially pancreatectomized animals. Blood glucose levels remained normal.

A 10% aqueous extract of young tender neem leaves fed to fasting rabbits at 200 mg/kg caused a marked fall in blood glucose concentration. A maximum reduction of 27% occurred 3 h after administration, and levels returned to normal after 12 h. A similar hypoglycemic effect was observed in fasted rats and, to a lesser extent, in guinea pigs following oral dosing.[47]

Neem seed is highlighted as a remedy for diabetes in India,[48] and "Nimbola" (a preparation based on the oil in gelatin capsules) is presently being manufactured by Kee-Pharma, B-34 Mayapuri Ind. Area, Phase-1, New Delhi, as an oral treatment for diabetes control.[49]

Intravenous injection of an aqueous solution containing 3 g of leaf extract into anesthetized dogs weighing 9 to 12 kg resulted in an initial rise in blood pressure followed shortly by a protracted fall, similar to the effect of histamine.[50]

E. MUSCULAR EFFECTS

Aqueous neem leaf extract caused an appreciable contraction of guinea pig ileum in Tyrode's solution.[50]

Injection of 0.5 to 1.0 mg of sodium nimbidinate into the rectus abdominis muscle of a frog failed to produce any effect on contraction caused by choline at 5×10^{-6} g/ml.[42] At 10^{-3} g/ml, the sodium salt produced a blocking effect at the myoneural junction. Administered intravenously, the sodium salt did not affect rabbit uterus *in situ* in doses of 10^{-5} to 10^{-2} g/ml, and 10^{-3} g/ml had no effect on nonpregnant rat uterus; 10 mg/kg administered intravenously to nonpregnant cats caused a relaxation of the uterine and vaginal muscles *in situ*. Injection of 100 μg of sodium nimbidinate into the artery supplying the dissected and perfused superior cervical ganglion of a cat, followed by electrical stimulation of the preganglionic fibers for 10 sec, failed to produce a contraction of the nictitating membrane; electrical stimulation at a higher strength and at longer duration likewise failed to produce transmission across the ganglion, showing that the salt had caused a blockage of the superior cervical ganglion. Intravenous injection of 10 μg of adrenalin produced a contraction of the ganglion.

F. GASTROINTESTINAL EFFECTS

Sodium nimbidinate at 10^{-6} g/ml applied for 30 sec to isolated intestinal strips of rabbit duodenum in Tyrode's solution lowered the tone without affecting the spontaneous activity.[42] A concentration of 10^{-3} g/ml caused relaxation of intestinal tone with little effect on spontaneous rhythmic activity. The same type of effect was noted when the drug was tested on frog intestine mounted in oxygenated Ringer's solution.

Nimbidin, a major bitter principle isolated from neem kernel oil, was administered orally as a colloidal solution in 10% alcohol to male albino rats at 20, 40, and 80 mg/kg, to female rats at 20, 30, 40, and 80 mg/kg, and to guinea pigs of either sex at 40 and 80 mg/kg.[51] All test animals were then subjected to stress with acetylsalicylic acid, serotonin, or indomethacin. Doses of 20 and 40 mg/kg gave significant protection from gastric and duodenal lesions in all animals. In ulcer-healing tests, nimbidin significantly enhanced the healing process in acetic acid-induced chronic gastric lesions in albino rats and dogs. Later tests showed that nimbidin is nontoxic to mice, rats, and dogs (see test results in Section III).

24-Methylenecycloartenol isolated from neem seed and dissolved in dimethylformamide at concentrations of 40 and 60 mg% inhibited the proteolytic activity of trypsin. Inhibition was greater with bovine serum albumin than with casein as the substrate.[52]

G. ONCOLYTIC EFFECTS

Chatterjee[53] reported cases in which an Indian patient with parotid tumors and another patient with epidermoid carcinoma were successfully treated with injections of neem seed oil. Hartwell[30] also reported that neem preparations were used to treat patients with various forms of cancer.

Evaluation *in vitro* of the neem limonoids, 7-acetylneotrichilenone and 1,2-diepoxy-azadiradione, against the murine P366 lymphocytic leukemia cell line showed that only the former was somewhat effective, with an ED_{50} of 8.5 μg/ml.[54]

H. NUTRITIVE EFFECTS

Considerable research has been conducted to determine the potential for the use of neem to supplement animal feeds.

1. In Poultry

Broiler chicks gained weight when fed a diet in which finely ground neem seeds or a hexane extract of the seeds was incorporated at 2.5, 5, or 7.5% as a replacement for groundnut cake. The largest weight gain was associated with a diet containing seed extract. Layer hens fed a diet in which the seed meal was replaced with 25, 50, 75, or 100% of the required nitrogen level showed reduced egg production and increased feed efficiency.[55,56]

For 6 weeks, 1-d-old broiler chicks were reared in thermostatically controlled battery jars on standard broiler starter mash, with or without 5 or 10% neem seed cake.[57,58] The birds fed untreated (control) diet showed the highest average weight gain. Birds fed a diet containing either concentration of neem seed gained little weight and showed 5 or 10% mortality, respectively.

2. In Sheep

Chokla lambs were fed a ration containing a mixture of 80% groundnut cake. 10% barley, and 10% molasses, and others were fed a ration in which 50, 75, or 100% of the groundnut protein was replaced with deoiled neem seed meal.[59] All diets were supplemented with minerals and vitamin A. Neem seed meal contained 17.65% crude protein and 25 to 42% crude fiber (dry matter basis). Palatability was poor among those animals fed neem-containing diet and several of these animals died after 28 d; surviving animals showed weight loss, diarrhea, severe gingivitis, sloughing of the mucous membrane of the tongue, and a foamy discharge from the mouth. Post-mortem showed severe congestion in the small intestine.[59] However, Gupta and Bhaid[60] reported that four 1-year-old sheep rams fed for 4 months on maize diets containing 25 or 50% deoiled neem seed cake, plus minerals and vitamins, showed no toxic symptoms. A diet containing 75% deoiled seed cake was consumed satisfactorily if the sheep were first adapted to a diet containing 25 or 50% neem cake. Severe weight loss resulted from maintaining sheep on a ration of 100% neem cake.

Vijjan et al.[61] studied the effect on male and female lambs of unextracted and alcohol-extracted neem seed cake, added to the normal ration as a substitute for 10, 20, or 30% of the wheat bran. Incorporation at the 10% level with extracted or unextracted seed cake had no effect on growth rate; 20% significantly increased the growth rate but 50% significantly reduced it. Blood glucose, hemoglobin, and urea nitrogen levels remained normal in all animals.

Chokla rams refused to feed on dried neem leaves mixed with molasses at the 10% level for 1 month. Dry leaves were also preserved in drums with and without minerals or molasses before feeding to rams each day for 2 weeks. Palatability was highest with mineral-enriched neem leaf ration, but all animals lost weight. Rams gained weight after feeding on a mixture of 60% barley, 20% groundnut cake, and 20% neem leaf meal for 50 d; feed consumption was good.[62]

3. In Livestock

Although it was known as early as 1941 that neem seed preparations were useful in veterinary medicine as applied to livestock,[63] it was not until the 1970s that the seed was successfully employed as a cattle feed additive.[64-68]

The digestibility of crude protein in neem seed cake for male buffaloes was 60.8 and 36.0% at 25 and 50% incorporation, respectively, in the diet. The concentrate mixture at 50% was unpalatable. Digestibility of various nutrients (dry matter, crude protein, ether extract, crude fiber, nitrogen-free extract, and total carbohydrates) at the 25% level was comparable to that of controls fed groundnut cake.[69] However, the palatability and growth rate of six cross-bred buffalo calves supplied 25% of their digestible crude protein as neem seed in the feed for 75 d were significantly lower than those of a control group fed regular diet.[70] In another group that was supplied with a diet containing 12.5% of seed cake for 100 d, the growth rate, intake, and digestibility of dry matter, crude protein, nitrogen-free extract, and total carbohydrates were all much lower.

Palatability and metabolism trials were conducted with 2-year-old buffalo calves fed iso-nitrogenous diets of starch or molasses containing 25 or 50% neem seed cake.[71] Although these diets proved to be unpalatable, the same concentrations of seed mixed with maize were palatable. However, digestive protein and nitrogen balances were poor and it was concluded that neem cake could not be considered as a true concentrate owing to its high fiber content.

Payne et al.[72] offered a normal ration containing 10, 15, or 20% neem seed cake to 5- or 6-year-old lactating Murrah buffaloes for 60 d beginning soon after calving. The fiber content of the cake was 25.84% (very high). All milk constituents tested (specific gravity, fat, protein, ash, lactose, and total solids) were normal, and there was no adverse effect on the health or feed intake of the animals. Blood serum protein values decreased with increasing content of neem cake, and cellular constituents of the blood were not affected.[73] Each animal was given 8 kg of treated paddy straw ration twice daily.

Lactating Murrah buffaloes were fed rations of groundnut cake, mustard cake, crushed gram, or wheat bran, with and without 10, 15, or 20% of neem seed cake on a protein-equivalent basis.[74] Each animal received 8 kg of the mixture per day plus 2 kg of mixed greens and unlimited amounts of paddy straw for 6 weeks. Palatability was satisfactory in all cases and total dry matter consumption and milk production were normal.

4. In Humans

It is said that the late Indian leader, Mahatma Gandhi, had a hearty respect for the nutritive value of greens. Learning that the leaves of the neem tree had extraordinary nutritional properties, albeit an incredibly bitter taste, he prepared a chutney of the leaves and ate it with gusto.[75] (Patrao[76] has very recently reported the discovery of a rare neem tree with sweet leaves.)

Fujiwara et al.[77,78] have studied in detail the water-soluble polysaccharides from neem bark which consist mainly of arabinofucoglucans, and Anderson et al.[79] found that the polysaccharide and proteinaceous components of the gum exudates include 18 amino acids.

III. TOXICOLOGY

A. FUNGI AND ALGAE

Aqueous extracts of neem seed cake were mixed (at unspecified concentrations) with potted field soils into which fungus-infested eggplant seedlings were introduced 15 d later. Parasitic fungi inhibited were *Fusarium* sp., *Rhizoctonia solani* Kuhn, and *Colletotrichus atramentarium* (Berk & Br.) Taubenh.[80,81] Tested conducted in petri dishes, liquid cultures, and on filter paper disks impregnated with an ethanolic extract of the seeds showed high effectiveness against plant-pathogenic fungi such as *C. papaya, Alternaria brassicae* (Berk.) Sacc., and *Helminthosporium* sp.[82] The liquid culture technique using an aqueous extract of the leaves gave maximum inhibition of mycelial growth of both species.

Germination of conidia of *Metarhizium anisopliae*, an entomopathogenic fungus infesting several important rice insect pests, was completely inhibited in a liquid saccharose-yeast medium containing 95% of neem oil expressed from seed; mycelium sporulation was also inhibited.[83] Although the effects were not so severe with 5 or 50% neem oil, significant inhibition was still obtained with a 5% concentration.

Khan and Wassilew[84] tested, on 14 species of common fungi pathogenic to humans, dry and fresh leaves and seeds as well as their extracts prepared with increasingly polar solvents (petroleum ether, methyl *tert*-butyl ether, chloroform, methanol, and water). All tests were conducted on culture medium contaminated with the fungus. Fungi used were *Trichophyton rubrum, T. violaceus, T. concentrichus, T. mentagrophytes, Epidermophyton floccosum, Microsporum gypseum, M. canis, Candida albicans, C. parapsilosis, Torulopsis glabrata, Trichosporon cutaneum, Scrophulariopsis brevicaulis, Geotrichum candidum,* and *Fusarium* sp. Unfortunately, the doses used were not given, but the antifungal effects decreased with increasing solvent polarity, indicating that the active component(s) are lipid in nature. The most effective extract was that of the leaves prepared with petroleum ether. Dermatophytes were especially affected.

Of 31 aqueous plant extracts tested against the fungus *Drechslera oryzae*, neem bark, leaves, and seeds were highly inhibitory when 10 spores were placed in each sterile paper disk which had been soaked in the extract.[85] Extracts were prepared by grinding 25 g of the plant part with 50 ml of water. The bark extract was most effective.

Neem seed cake increased the populations of blue-green algae and their nitrogen-fixing ability in beakers and in soil in the field at 57 kg/ha.[86]

Nimbidin and thionimone, two neem seed components, were tested as 0.1, 1.0, and 5% solutions in dilute ethanol in petri dishes on the growth of two isolates of *Rhizoctonia solani, Fusarium oxysporum* f. *lycopersici* Schlecht., *Helminthosporium nodulosum, Alternaria tenuis* Aust., and *Curvularia tuberculata*. All concentrations were highly inhibitory, the growth decreasing rapidly with increasing concentration.[81]

B. BACTERIA AND MUTAGENICITY

The standard Ames test conducted with 100 to 3000 μg of azadirachtin, the major insecticidal component of neem seeds, and another component, salannin, showed no mutagenic activity on four strains of *Salmonella typhimurium*.[87-90] Nimbolide and nimbic acid were also negative in TA98 and TA100 strains of this organism,[91,92] as was extracted neem seed oil.[93] Nimbolide, at 0.875 mg impregnated on paper disks, showed antibacterial activity on 3 of 17 strains of *Staphylococcus aureus, S. coagulase* (+), and *S. coagulase* (−). Nimbic acid at 3.5 mg/disk exhibited this activity on *S. aureus, Bacillus subtilis, S. coagulase* (+ and −), and Diphtheroidae. No antibacterial activity was exhibited against 11 bacterial species (*Citrobacter*, 2 strains of *Escherichia coli*, *Enterobacter, Klebsiella pneumoniae, Proteus mirabilis, P. morgasi, P. vulgaris, Pseudomonas aeruginosa, Pseudomonas* E01, and *Streptococcus faecalis*).[92]

Neem leaves were extracted with solvents of increasing polarity and the extracts were tested against *Staphylococcus aureus*.[93,94] The most effective extract, prepared with methyl *tert*-butyl ether, was purified by silica gel chromatography using a stepwise gradient from ethyl acetate to methanol. Further purification by gel chromatography gave a mixture of compounds that suppressed the organism at 1000 ppm.

C. VIRUSES AND PROTOZOA

Singh[95] found that potato X-virus could be completely inactivated by mixing an alcoholic extract of neem seeds with equal amounts of distilled water and a standard inoculum of the virus. Verma[96] identified nimbin and nimbidin as the components responsible for this inactivation 3 years later.

Several European and Indian physicians in the 19th century recommended the use of neem leaves and bark as a febrifuge, especially in cases of malaria.[35] However, in 1976 Tella[97] determined by clinical tests that an aqueous extract of the leaves, administered orally to young male albino mice in which primary acute blood-transmitted *Plasmodium berghei* malaria had been induced, failed to affect the course of the disease. It is, of course, possible that any components in the leaves or the bark that may be effective against this plasmid are not water-soluble. Inasmuch as malaria is such a prevalent and debilitating disease in many parts of the world, further investigation of various parts of the neem tree against this disease is indicated.

D. INVERTEBRATES

1. Aquatic Organisms

a. Snails

An ethanolic extract of neem seeds was toxic to the snail, *Biomphalaria glabrata* (Say), and to its eggs in laboratory tests.[98,99] These findings are in line with those reported earlier using "neem margosa" against the vector snail, *Melania scabra*.[100]

The freshwater snail, *Limnea natalensis*, is a serious azolla pest in Philippine rice fields and a vector for schistosomiasis. A mixture of 1.5 g of neem seed cake and 1.5 g of metaldehyde failed to control the snail in laboratory azolla inoculum.[101]

b. Shrimp

No mortality resulted among brine shrimp, *Artemia salina* Leach, kept in contact with paper disks moistened with 4 ml of ethanolic or hexane extracts of neem seeds for 24 h. Contact for 48 h was not lethal with the hexane extract but the ethanolic extract killed 18% of the shrimp.[102]

c. Ostracods

Ostracods are crustaceans that feed actively on nitrogen-fixing blue-green algae. The effects of 10 and 3.2 ppm of an aqueous neem seed extract on first and second instars of the ostracod, *Heterocypris lugonensis* Neale, were determined in submerged tropical soils as well as under laboratory conditions.[103,104] Ostracod populations were suppressed in all soils except those highly organic in nature. Reduction of growth in the laboratory occurred 2 and 5 d after treatment, respectively, at these concentrations. All animals exposed to 10 ppm died within 11 d. No morphological damage resulted from treatment at 0.32 or 1 ppm. Advanced instars (seventh and eighth stages) exposed to the extract at concentrations of 1 to 100 ppm laid eggs that hatched, but the resulting young displayed severe valve deformation. Over sand, concentrations above 1 ppm interfered with molting.

2. Earthworms and Nematodes

In greenhouse experiments, ground neem leaves or seed kernels were mixed with soil at 5% concentration to study their effect on the earthworm *Eisenia foetida*. Earthworms in treated pot soils showed significantly greater weight and lower mortality compared to worms in untreated soil 4 weeks after application. Development was greatest in soil containing ground neem kernels in the upper third of the substrate. In the field, however, no significant difference was seen in the numbers of animals in treated or untreated soils, although the average weight of the individual worms was highest in treated plots.[105]

Various parts of the neem tree have been tested extensively for their effects on a number of species of nematodes. Although a good review of this subject has been published by Jain[106] and more recently by Vijayalakshmi et al.,[107] these are not readily available and a more detailed review is therefore provided here.

Aqueous suspensions of ripe neem berries were not effective in controlling nematodes of an unidentified species of *Meloidogyne* when applied (dose not given) to the soil in Gambia.[108] The degree of infestation, as assessed by the abundance of root-knot nodules, was not consistent with control treatment or between replicates. However, the limonoid bitter fraction obtained following extraction of the oil from the berries with hexane was quite effective in killing both the nematode eggs and adults in India[109] and in the Philippines.[110]

The water-soluble portions of the leaves, fruits, and seed cake of neem, used as soil amendments, reduced the population of a wide variety of plant-parasitic nematodes.[81,111] The seed oil inhibited and killed the larval stages of the root-knot nematode, *Meloidogyne incognita* (Kofoid & White) Chitwood.[81] Aqueous extracts of the seed gum and, especially, of the leaves were highly toxic to this species as well as to the reniform nematode, *Rotylenchus reniformis* Linford and Oliveira, in laboratory tests.[112,113] This species was more sensitive than *M. incognita*.

Aqueous extracts of neem leaves, seeds, and flowers tested at 10, 5, and 1% in petri dishes completely prevented hatching of *M. incognita* eggs; concentrations of 1, 0.1, and 0.001% were also toxic but required a longer period of contact.[114] The seed extract and an extract of the fruit were especially active.[115] In contrast, Johnson[116] reported that the application of 0.2 or 0.4% solution of seed extract to the soil in which flue-cured tobacco plants were growing in the field failed to suppress this species of nematode within 3 months or root galls within 5 months. Neither plant growth nor yield was improved by the treatment.

In an attempt to identify the active nematicidal principles, Devakumar et al.[117] fractionated an alcoholic extract of the seed kernels into the oil and the less polar and more polar limonoids. The toxicity to egg hatch and to second stage *M. incognita* larvae was concentrated in the less polar fraction, although some activity was shown by the more polar limonoids; the oil was completely ineffective.

Extracts of neem seed cake prepared with hot water or with methanol were highly toxic to *M. incognita* larvae.[118,119]

Neem leaves and seeds were powdered and 10 g of each was ground separately with 50 ml of distilled water. The resulting extract was diluted to 5%, transferred to petri dishes at 5 ml/dish, and *M. incognita* egg masses were placed in each dish. Observations made at 3-d intervals for 30 d showed a reduction in egg hatch of at least 37% and a high level of larval mortality for each extract.[120]

Roessner and Zebitz[121] found that the growth of healthy tomatoes in potted soil containing 1% of neem seed kernels was increased almost 50% due to the deleterious effect of neem on galls of *Meloidogyne arenaria* Chitwood; the substitution of neem leaves for kernels was somewhat less beneficial. Similar results were obtained with galls of *Pratylenchus penetrans* Filipjev & Sch. Stekhoven. Root-knot of vegetables caused by *M. javanica* Chitwood was severely reduced by amending the nutrient soil with 0.1 to 2.0% of neem seed oil or extracts of the seed.[122] Results varied with the nematodes *Haplolaimus indicus* Sher, *Tylenchorhynchus indicus*, and *Aphelenchus avenae* Bestian.

Consistent increases in vegetable yields and considerable reductions in nematodes of the genera *Meloidogyne* and *Rotylenchulus* resulted from soil amendment with neem cake.[123-126] The cake was mixed with field soil in pots containing eggplant seedlings infested with the nematodes. Although 38 species were not affected by neem,[80] *M. incognita, R. brassicae, R. reniformis, Helicotylenchus erythrina* Golden, and *Haplolaimus indicus* were strongly suppressed. Application of 300 kg of neem leaves per hectare into the soil was very effective in reducing the numbers of species of *Meloidogyne, Tylenchorhynchus,* and *Helicotylenchus* infesting grape vines.[127]

The ectoparasitic soil nematodes *Haplolaimus indicus, Helicotylenchus indicus,* and *Tylenchus filiformis* Butschli kept in petri dishes with aqueous extracts of neem leaves, flowers, or fruits were keenly affected after 48 h. Maximum toxicity was caused by 1% fruit extract, followed by 10% leaf and flower extracts. Neem bark extract at 10% was also effective.[128]

Petri dish tests were conducted using various dilutions (60, 90, and 350 times) of an aqueous extract of neem leaves, together with several species of nematodes.[129] These diluted concentrations caused 50% mortality of *Tylenchus brassicae* Siddiqi within 12, 24, and 48 h, respectively. Comparative figures were 45, 55, and 160 h for *T. filiformis* and 50, 66, and 300 h for *Helicotylenchus indicus*.

Egunjobi and Afolami[130,131] obtained drastic reductions of *Pratylenchus brachyurus* Filipjev & Sch. Steckhoven populations in maize roots by inoculating the soil with 0.5, 1.0, or 1.5 kg of aqueous neem leaf extract per 3 l of water. Root populations of nematodes 11 weeks following inoculation at these doses were 29.8, 48.2, and 70.8%, respectively. Maize yield and root weights were negatively correlated with root nematode populations.

Gill and Lewis[132] were probably the first investigators to determine that neem was detrimental to *P. brachyurus*. When barley seedlings were dipped in a 1% aqueous suspension of crushed kernels for 2 h and then grown for 10 d in soil infested with this nematode, the number of nematodes found in the root system was reduced by more than 50%. In all probability this observation was also the first substantiated report of the systemic action of neem in certain species of plants.

Kaplan[133] conducted a 2-year study in Florida to determine the effect of an ethanol extract of neem kernels on several species of nematodes. Potted citron (*Citrus indicus*) plants in a greenhouse were kept in soil inoculated with 100, 500, or 5000 nematodes/pot. After inoculation (2 months) the trees were treated with 0 (control), 1000, 2000, or 4000 ppm extract/200 ml of water as a soil drench. Population densities, total fibrous root weight, and stem diameter were determined 6 months later, and it was found that *Radopholus citrophilus* Thorne, *Pratylenchus coffeae* Filipjev & Sch. Steckhoven, and *Tylenchulus semipenetrans* Cobb had suffered no adverse effect.

The plant parasitic nematode, *Xiphinema basari* Siddiqi, is a major pest of fruit trees in the Sudan. It occurs fairly prominently in association with the roots of neem trees. Greenhouse tests showed the nematode to be highly parasitic to neem seedlings in potted soil, resulting in stubbled and stunted roots.[134-136]

E. VERTEBRATES
1. Fish and Frogs
"Neem oil extractive", a waste product from neem oil refining, was an effective larvicide for *Culex* mosquitoes at 0.005% as an aqueous emulsion and was nontoxic to insectivorous fish, *Gambusia* sp.[137] However, 100% of the fish were killed in 24 h by a concentration of 0.04%; this concentration also killed more than 80% of tadpoles in 24 h and 100% in 2 d, although the extractive was not toxic to the tadpoles at 0.01%.

2. Birds
Ketkar[138] reported that birds eat the sweet outer pulp and skin of neem fruit but spit out the seed, which has a bitter taste. El Amin et al.[139] reported that the fruits are eaten by the white-vented bulbul bird, *Pycnonotus barbatus* (Desf.), as well as other Pycnonotidae. According to Doria[140] and Kubo and Nakanishi[141] neem berries are the favorite fruit of certain tropical birds in Kenya and Nigeria. A study conducted by Sengupta[142] in India indicated that house sparrows add neem leaves to their nests, ostensibly to protect them from parasites.

Despite these reports, Singh et al.[143] found that an aqueous extract of the ripe berries (100 berries boiled in 1 l of distilled water for 6 h, filtered, and concentrated to 200 ml) was quite toxic to 3- to 4-month-old White Leghorn chickens. The extract, equivalent to 5 g of berries, was administered orally as a single dose. Shortly thereafter, the treated birds appeared sluggish, stopped feeding, and exhibited drooping head and cyanosed combs; 60% of the birds died within 24 h. Post-mortem showed fragile livers with congestion, bile retention, and congested hemorrhaging kidneys.

An ethyl ether extract of neem seed oil failed to repel red-winged blackbirds, *Agelaius phoeniceus*, when tested in an aviary under no-choice conditions for 18 consecutive hours by the procedure of Starr et al.[144] The amounts of treated food consumed were 81.6 and 68.8%, respectively. For acute oral avian tests, the birds were dosed by gavage with known levels of expressed and extracted oil in propylene glycol; two individual caged birds were used for each dose level. The LD_{50} was above 1000 mg/kg for expressed oil and 1000 mg/kg for extracted oil.[145]

Trial feeding of neem seed meal to starter chickens resulted in severe hepatitis with necrotic patches, mild to severe nephritis with congestion and slight inflammation and scattered petechiae in the intestine.[96] However, it is highly likely that contamination of the seed with aflatoxin was responsible for these effects, which are similar to those reported with aflatoxicosis in chickens.[146]

3. Rodents
Laboratory mice of mixed sexes weighing 23 to 30 g received single oral doses of 1.6, 3.2, 6.4, 8.5, 10.6, and 12.8 g/kg of a methanolic (75%) extract of neem leaves and bark. All doses were administered as 2 ml of a solution of the extract in 2% sodium carbonate solution. The acute oral toxicity was approximately 13 g/kg, with the mice showing discomfort, gastrointestinal spasms, apathy, and hypothermia.[39,147]

Vijjan and Tandan[148] studied the effect of oral treatment of adult albino mice with an aqueous extract of neem seed cake (0.5 and 1.0 g/kg body weight) for 3 consecutive days on pentobarbitone-induced sleeping time. Animals were administered sodium pentobarbitone (40 mg/kg) intraperitoneally 48 h after the last dose and the duration of sleeping time was recorded. Sleeping time in the controls (given the drug only) and those animals receiving 0.5 and 1.0 g/kg of neem extract was 59.3 ± 8.5, 83.0 ± 12.6, and 170 ± 19.9 min, respectively, indicating that the extract may interfere with the hepatic microsomal enzyme system to prevent pentobarbitone breakdown.

Bhide et al.[149] tested the acute toxicity of sodium nimbidinate, obtained from the bitter fraction of the seed oil, in mice by dissolving it in distilled water and injecting it intraperitoneally; the LD_{50} was 700 mg/kg body weight. The acute toxicity (LD_{50}) when given orally to rats was 1000 mg/kg. Histological changes included cloudy liver and renal tubules, necrotic changes in the tubules, and congested glomeruli due to the proliferation of endothelial cells.

Pillai and Santhakumari[150] conducted acute and subacute toxicity tests with nimbidin by oral and intraperitoneal administration to mice and rats. Oral administration was given with 20 to 200 mg/kg suspended in gum tragacanth and 10% alcohol; intraperitoneal administration was at 20 to 1000 mg/kg. Subacute toxicity was evaluated only in albino rats by single daily oral administration of 25, 50, or 100 mg/kg for 6 weeks. No toxicity was evident in any of these tests, and teratogenic studies in rats did not reveal any toxic manifestations or fetal abnormalities.

Neemrich-100 (technical grade) is a formulation of 30% oil extracted from dry neem seeds. A requirement for the registration of this preparation was the determination of the effect of repeated dermal application to rats. Male and female albino rats (Wistar strain) were painted on the shaved skin with the formulation at rates of 200, 400, or 600 mg/d on 5 d/week for 3 weeks, kept under observation for an additional 2 weeks, and then sacrificed. Although no clinical symptoms of poisoning were seen, the skin had become roughened and thick, especially at the highest dose. Livers of rats tested at the highest dose showed a 40% increase in size, but others showed no abnormality. Microscopic examination of skin sections after treatment revealed scaling of the epidermis and hyperkeratosis of the stratum corneum; severity increased with increasing dosage. Histopathological evaluation of the lungs, kidneys, and testes showed no difference between treated and untreated animals. Treated rats exhibited higher food consumption, gained weight, and showed no abnormal blood levels.[151]

Oral and dermal administration of neem seed cake (dosages not given) to rats caused no carcinogenicity.[152]

Aqueous suspensions of green or dried neem leaves administered orally to guinea pigs at 50 or 200 mg/kg for up to 8 weeks caused a progressive decrease in body weight and pulse and respiratory rates, as well as diarrhea in those animals fed the fresh leaves.[153]

4. Rabbits

No reactions were observed within 6 d after 0.1 mg of an ethanol extractive of neem seeds was instilled into one eye of a number of male albino New Zealand rabbits.[89] Solutions (1 to 5%) of sodium nimbidinate instilled into the eyes failed to cause any eye irritation and there was no change in the size of the pupil. No local irritation or induration resulted from subcutaneous and deep intramuscular injection at 60 mg/kg.[42] Instillation of a 10% solution caused no irritation.[149]

5. Sheep and Goats

A 2-year-old male sheep allowed to feed on 100 g of fresh neem leaves showed nervous responses of the head, circular movements, slight tympanism, dyspnea, and an increase in body temperature.[154] These symptoms lasted for 12 h before the animal died. No post-mortem was performed.

Nubian goats 1 to 2 years old fed a drench of green or dried neem leaves at 50 or

200 mg/kg/d for up to 8 weeks exhibited general weakness, decreased heart and respiratory rates, and diarrhea, but no hematological changes.[153] Necropsy showed areas of hemorrhagic erosion, congestion and degeneration of the liver, kidneys, lungs, duodenum, brain, and seminiferous tubules.

6. Dogs

Bhide et al.[149] fed gradually increasing doses (up to 200 mg/kg) of sodium nimbidinate to dogs. These dogs were also injected both intramuscularly and intravenously with the same doses. The maximum intravenous dose was 2400 mg/kg. Adult female dogs (10 and 13.5 kg, respectively) received 60 mg of the compound intramuscularly each day for 60 d and were then sacrificed for microscopic tissue study. Although no toxic manifestations were seen during the course of the experiments, six puppies delivered from a large female shortly thereafter died within 10 d "for no apparent reason". The kidneys of all puppies showed cloudy degeneration of the tubules.

Mongrel dogs given a single dose of 10 to 20 mg of nimbidin/kg in 10% alcohol by intubation each day for 28 d showed no toxic effects. Necropsy findings and histological examination of the liver, kidneys, and heart were normal. Hematological and biochemical analyses revealed no systemic toxicity.[150]

7. Humans

Caius and Mhaskar (as reported by Chopra et al.[155]) found that neem seed oil produced occasional diarrhea, nausea, and general discomfort when given orally as an anthelmintic. The leaves and oil proved to be ineffective in expelling intestinal parasites.

Sodium nimbidinate fed in 7-g doses to human subjects or injected intramuscularly at 1 g elicited no local or general side effects.[149,156]

Within hours of ingestion, 13 Malaysian infants given large (up to 5 ml) oral doses of neem seed oil for minor ailments developed vomiting, drowsiness, metabolic acidosis, and encephalopathy.[157] A 4-month-old Indian infant was given 12 ml of the oil on each of 2 successive days for cough. The second dose was followed within 1.5 h by vomiting, drowsiness, respiratory difficulty, and seizures. The liver became enlarged and the child died 12 d after ingestion of the oil.[158] These findings indicate that the oil may be involved in the etiology of Reye's syndrome, which in turn may be caused by a synergistic effect of aflatoxin and other toxic components present in the oil.[159]

A minireview of the pharmacological and toxicological effects of neem on warm-blooded animals, with 40 references, has recently been published by Jacobson.[160]

For the toxicological data on a commercial insecticidal formulation of neem seed extract named "Margosan-O" see Chapter 9.

IV. ANTIFERTILITY EFFECTS

In 1959, Sharma and Saksena[161,162] published the first reports of possible antifertility (spermicidal) activity by neem components in rats and humans. Numerous subsequent reports by other investigators have largely substantiated this activity in humans and other animals in both *in vitro* and *in vivo* research, culminating in the recent commercialization of a neem formulation for use as a human contraceptive. The developments have proved to be so interesting and significant, not only to Indian birth control programs but also for the potential use of these and similar products worldwide, especially in overpopulated Third World countries, that a separate section on this subject seems appropriate.

Using the method of Baker et al.,[163] Sharma and Saksena[161,162] conducted *in vitro* tests of the effects of sodium nimbidinate on rat and human sperm. A mixture of 2 ml of semen, 1 ml of air, and 2 ml (10 mg%) of an aqueous solution of the salt was used. Untreated (control) rat suspension showed 2.5 and 10% dead sperm after 5 and 30 min, respectively. A concentration of 50 mg% killed all rat sperm in 5 min; 2.5 mg% required 30 min. Control human semen showed highly motile sperm after 5 and 30 min, whereas the mixture of semen plus the sodium salt at 1000 and 250 mg% showed all dead sperm in 5 and 30 min, respectively. A concentration of 500 mg% killed 14.16% of human sperm. These results indicate the absence of close correlation between rat and human sperm.[164]

Deshpande et al.[165,166] homogenized 50 g of crushed green neem leaves with 100 ml of water, filtered the mixture, and fed 1 ml of the solution each day for 1 month with standard diet to male and female mice. Animals were then either sacrificed or mated. Whereas control animals showed a 100% fertility rate, treated females that were mated showed reduced pregnancies and litter size. Males showed normal reproductive function 1.5 months following treatment, indicating that the antifertility effect in the male mouse is reversible and spermatogenesis is not involved.

In 1983, Sadre et al.[167] reported on a continued study of the effect of neem leaf aqueous extract, prepared from 100 g of leaves in 200 ml of water, on the fertility of rats, rabbits, and guinea pigs. Male rats fed a standard diet plus a daily dose of 2 ml of the test solution showed a 66.7% reduction in fertility after 6 weeks, 80% after 9 weeks, and 100% after 11 weeks, with no inhibition of spermatogenesis, no decrease in body weight, and no toxic manifestations. However, a marked decrease in sperma-tozoal motility was observed, unassociated with loss of libido or impotence. The infertility

In 1983, Sadre et al.[167] reported on a continued study of the effect of neem leaf aqueous extract, prepared from 100 g of leaves in 200 ml of water, on the fertility of rats, rabbits, and guinea pigs. Male rats fed a standard diet plus a daily dose of 2 ml of the test solution showed a 66.7% reduction in fertility after 6 weeks, 80% after 9 weeks, and 100% after 11 weeks, with no inhibition of spermatogenesis, no decrease in body weight, and no toxic manifestations. However, a marked decrease in sperma-tozoal motility was observed, unassociated with loss of libido or impotence. The infertility was reversible in 4 to 6 weeks. When male guinea pigs were given 2-ml doses of the solution orally, 66.6 and 74.9% mortality resulted after 4 and 6 weeks, respectively. Male rabbits given 5 ml of the solution/kg of body weight showed 80 and 90% mortality after 4 and 6 weeks, respectively.

Female albino rabbits fed a standard diet were given 2, 4, or 6 ml of neem seed oil/ kg orally for 3 d prior to inducement of ovulation by means of copper acetate injection. Mild antiovulatory activity was seen with all doses. Pregnant female rats given the oil for 18 d at the same concentrations showed significantly lengthened dioestrus phase (mild antifertility activity).[168] Male rats fed the oil in graded doses for 1 month showed a remarkable loss of reproductive function insofar as impregnation of females was concerned. The number of pregnancies again reached normalcy following a 1-month oil-free interval. Although inhibition of spermatogenesis was not seen in treated males, some of these animals receiving the oil at 6 ml/kg showed temporary signs of decreased motor activity and loss of appetite, which reverted to normalcy with cessation of treat-ment. It is speculated that the oil induces functional infertility in male rats without in-terfering with the structural integrity of the testes or the process of spermatogenesis, although it may interfere with sperm transportation into the epididymis and passage through the vas deferens.[168]

Krause and Adami[169] confirmed that neem seed kernel oil extracted with methanol did not inhibit spermatogenesis in male rats fed daily doses of 0.1 ml of a 10% extract (diluted with water to 1%) for 1 to 9 weeks. Total weights of the animals and of their testes, epididymis, and seminal vesicles, as well as testosterone levels in the blood, remained normal.

Neem oil purchased on the open Indian market was shown to possess strong *in vitro* spermicidal action (within 30 s) with monkey and human semen.[170] When instilled intravaginally at 20 µl in rats and at 1 ml in monkeys and human subjects, the oil was completely effective in preventing pregnancy without causing side effects, as confirmed by histopathological studies. The odor of the oil, found to be unpleasant by several couples in the study, was satisfactorily masked by adding 0.5 ml of lemon grass scent to 100 ml of the oil. Menstrual cycles of the women maintained regularity with no bleeding abnormality. These same investigators undertook a study to establish an anti-implantation effect of neem oil and its aftereffects.[171,172] Single or multiple intravaginal applications of 25 or 75 µl of the oil into rats completely prevented implantation; 12.5 µl inhibited implantation by 50%. Three consecutive days of vaginal application was required for optimum anti-implantation effect, although a single application of a large dose was effective. Histological examination of the ovaries and the cervical end of the uterine horns showed no abnormal changes. Following withdrawal of oil application for 3 d, 30% restoration of fertility occurred; the resulting rat pups showed no abnormality.[172]

Intravaginal administration of 0.1 ml of neem oil/d for the first 10 d of pregnancy in albino rats resulted in smaller fetuses by day 16, indicating resorption.[173] Oral administration of 0.1 ml of oil/d for the first 10 d of pregnancy caused 40% antifertility, with 50% of the animals showing smaller fetuses. Of those animals treated orally and allowed to complete the pregnancy, 80% failed to deliver.

The effect of neem oil on implantation in pregnant female rats was investigated by Tewari et al.[174] Mating was confirmed by the presence of a vaginal plug and spermatozoa in smears, which was considered as day 1 of pregnancy. Various doses of oil expressed from crushed seed were administered subcutaneously for various numbers of days after coitus. Concentrations of 0.05 to 0.30 ml/rat administered 5 or 7 d after coitus showed significant antifertility activity. A dose of 0.2 ml inhibited pregnancy in 100% of the rats when given at the same period, but no significant anti-implantation activity was observed on a 3-d schedule. Oral administration of 2.5 and 5.0 ml oil/kg/d did not cause significant antifertility activity and ovulation occurred normally. This study further corroborated the anti-implantation effect of neem oil by the subcutaneous route.

The contraceptive effects of neem oil demonstrated in higher animals and in humans is not altogether surprising in view of the fact that the oil has reduced the fecundity of and sterilized a number of insect species; e.g., in the rice moth [*Corcyra cephalonica* (Stainton)],[175] the green rice leafhopper [*Nephotettix virescens* (Distant)],[176] the Colorado potato beetle (*Leptinotarsa decemlineata* Say),[177] and the ant, *Formica polyctena* Foerster.[178]

The results obtained thus far concerning the nontoxic and contraceptive effects of neem oil are highly encouraging and clearly warrant additional research with other species of higher animals and man. Determination of the active principles in the oil responsible for the antifertility effects would also be of considerable interest, and research directed toward this end is presently in progress.[172] In the meantime, it is encouraging to see that a formulation based on neem oil is being manufactured and sold by Excelsior Enterprises, B-12, A. M. Jaipuria Road Cantt., Kanpur-208004, U. P. India for insertion into the woman's vagina prior to coitus.[179] This preparation, labeled "Sensal", is claimed to be also an antiseptic and a vaginal toner, with "no contraindications reported".

ACKNOWLEDGMENT

I thank Dr. B. S. Parmar, Editor of Neem Newsletter, Division of Entomology, Indian Agricultural Research Institute, New Delhi, India, for permission to use small amounts of data published in my previous minireview in *Neem Newsletter* in 1986.[160]

REFERENCES

1. **Radwanski, S. A.,** Neem tree. I. Commercial potential, characteristics, and distribution, *World Crops Livestock,* 29, 63, 1977.
2. **Duke, J. A.,** personal communication, 1981.
3. **Radwanski, S. A.,** Neem tree. II. Uses and potential uses, *World Crops Livestock,* 29, 111, 1977.
4. **Radwanski, S. A.,** Neem tree. III. Further uses and potential uses, *World Crops Livestock,* 29, 167, 1977.
5. **Lewis, W. H. and Elvin-Lewis, M. P. F.,** *Medical Botany,* John Wiley & Sons, New York, 1977, 47, 369.
6. **Steyn, D. C. and Rindl, M.,** Preliminary report on the toxicity of the fruit of *Melia azedarach* (syringa berries), *Trans. R. Soc. S. Afr.,* 17, 295, 1929.
7. **Morrison, F. R.,** Contribution on the chemistry of the fruit obtained from the white cedar tree (*Melia azedarach* L. var. *australasica* C. DC: syn. *Melia australasica* A. Juss) growing in New South Wales, with notes on its reputed toxicity, *Proc. R. Soc. New South Wales,* 65, 153, 1932.
8. **Morton, J. F.,** *Plants Poisonous to People in Florida and Other Warm Areas,* 2nd ed., Southeastern Printing Co., Stuart, Fla., 1982, 35.
9. **Oelrichs, P. B., Hill, M. W., Vallely, P. J., MacLeod, J. K., and Molinski, T. F.,** Toxic tetranortriterpenes of the fruit of *Melia azedarach, Phytochemistry,* 22, 531, 1983.
10. **Warthen, J. D., Jr.,** *Azadirachta indica:* a source of insect feeding inhibitors and growth regulators, U.S. Department of Agriculture, Agric. Rev. Manuals, ARM-NE4, 1979.
11. **Radwanski, S. A. and Wickens, G. E.,** Vegetative fallows and potential value of the neem tree (*Azadirachta indica*) in the tropics, *Econ. Bot.,* 35, 398, 1981.
12. **Radwanski, S. A.,** The Uses, Ecology and Silviculture of *Azadirachta indica:* Neem Tree, unpublished data, 1980.
13. **Ahmed, S. and Grainge, M.,** Potential of the neem tree (*Azadirachta indica*) for pest control and rural development, *Econ. Bot.,* 40, 201, 1986.
14. **Jacobson, M.,** The neem tree: natural resistance par excellence, *Am. Chem. Soc.*

15. **Jacobson, M.,** Natural pesticides, in *Research for Tomorrow,* Crowley, J. J., Ed., 1986 Yearbook of Agriculture, Washington, D.C., 1986, 144.
16. **Fleming, J. F.,** *A Catalogue of Indian Medicinal Plants and Drugs,* Hindustani Press, Calcutta, India, 1810.
17. **Waring, E. J.,** *Pharmacopoeia India,* Allen, London, 1868.
18. **Dymock, W.,** *The Vegetable Materia Medica of Western India,* Trubner, London, 1885.
19. **Dutt, V. C.,** *The Materia Medica of the Hindus,* Calcutta, India, 1910.
20. **Mitra, C.,** Investigations on neem (*Melia indica*) and its oil, *Indian Oilseeds J.,* 1, 256, 1957.
21. **Mitra, C. R.,** Neem. India Central Oilseeds Commission, Hyderabad, 1963.
22. **Das, S. K.,** *Medicinal, Economic and Useful Plants of India,* Prachi Gobeson, Calcutta, India, 1975, 56.
23. **Jotwani, M. C. and Srivastava, K. P.,** Neem insecticide of the future. III. Chemistry, toxicology and future strategy, *Pesticides,* 15(12), 12, 1981.
24. **Nadkarni, K. M. and Nadkarni, A. K.,** *Indian Materia Medica,* Popular Prakashan, Bombay, India, 1976.
25. **Hoeg, O. A.,** Twigs from *Azadirachta* used as toothbrushes in India, *Blyttia,* 33, 125, 1975.
26. **Anon.,** New toothpaste ingredient, *Chem. Drug.,* 200, 125, 1973.
27. **Ahmed, S.,** Utilizing Indigenous Plant Resources in Rural Development: Potential of the Neem Tree, East-West Center, University of Hawaii, Honolulu, 1985.
28. **Ganesalingan, V. K.,** The use of neem (*Azadirachta indica*) in Sri Lanka, Abstr. 3rd Int. Neem Conf., Nairobi, Kenya, July 1986, 46.
29. **Lal, S. D. and Yadav, B. K.,** Folk medicines of Kurukshetra District (Haryana), India, *Econ. Bot.,* 37, 299, 1983.
30. **Hartwell, J. L.,** *Plants Used Against Cancer, a Survey,* Quarterman, Lawrence, Mass., 1982, 33, 181.
31. **Drury, H.,** *The Useful Plants of India,* Asylum Press, Madras, India, 1858.

Symp. Ser., 296, 220, 1986.

32. **Kaviratna, A. C.,** *Charaka-Samhita,* Vols. 1 and 2, Corinthian Press, Calcutta, India, 1888—1909.

33. **O'Shaughnessy, W. H.,** *The Bengal Dispensary and Pharmacopoeia,* Bishop's College Press, Calcutta, India, 1841.

34. **Lewis, W. H. and Elvin-Lewis, M. P. F.,** Neem (*Azadirachta indica*) cultivated in Haiti, *Econ. Bot.,* 37, 69, 1983.

35. **Siddiqui, S. and Mitra, C.,** Utilization of nim oil and its bitter constituents (nimbidin series) in the pharmaceutical industry, *J. Sci. Ind. Res., India,* 4, 5, 1945.

36. **Patrao, H.,** Neemcure, *Neem Newslett.,* 1, 50, 1984.

37. **Duke, J. A. and Wayne, K. K.,** *Medicinal Plants of the World,* 3 vols., Computer Index, 1981.

38. **Anon.,** Doctors find neem good for skin diseases, *New Delhi Evening News,* January 29, 1985.

39. **Okpanyi, S. N. and Ezeukwu, G. C.,** Anti-inflammatory and antipyretic activities of *Azadirachta indica, Planta Med.,* 41, 34, 1981.

40. **Radwanski, S. A.,** personal communication, 1984.

41. **Bhide, N. K., Mehta, D. J., and Lewis, H. A.,** Diuretic action of sodium nimbidinate, *Indian J. Med. Sci.,* 12, 141, 1958.

42. **Gaitonde, B. B. and Sheth, U. K.,** Pharmacological studies of sodium nimbidinate, *Indian J. Med. Sci.,* 12, 156, 1958.

43. **Pillai, N. R. and Santhakumari, G.,** Hypoglycaemic activity of *Melia azadirachta* Linn (neem), *Indian J. Med. Res.,* 74, 931, 1981.

44. **Garg, H. S. and Bhakuni, D. S.,** 2',3'-Dehydrosalannol, a tetranortriterpenoid from *Azadirachta indica* leaves, *Phytochemistry,* 24, 866, 1985.

45. **Thompson, E. B. and Anderson, C. C.,** Cardiovascular effects of *Azadirachta indica* extract, *J. Pharm. Sci.,* 67, 1476, 1978.

46. **Bhargava, A. K., Dwivedi, S. K., and Raj Singh, G. A.,** A note on the use of neem oil (*Azadirachta indica*) as antihyperglycaemic agent in dogs, *Indian J. Vet. Surgery,* 6, 66, 1985.

47. **Luscombe, D. K. and Taha, S. A.,** Pharmacological studies on the leaves of *Azadirachta indica, J. Pharm. Pharmacol.,* 26 (Suppl.), 110P, 1974.

48. **Anon.,** Neem oil as a remedy for diabetes, Amar Ujala, Bareilly, India, July 7, 1985.

49. **Anon.,** Nimbola, *Neem Newslett.,* 3, 37, 1986.

50. **Arigbabu, S. O. and Don-Pedro, S. G.,** Studies on some pharmaceutical properties of *Azadirachta indica* or baba yaro, *Afr. J. Pharm. Pharmaceut. Sci.,* 114, 181, 1971.

51. **Pillai, N. R. and Santhakumari, G.,** Effects of nimbidin on acute and chronic gastro-duodenal ulcer models in experimental animals, *Planta Med.,* 50, 143, 1984.

52. **Banerjee, R. and Nigan, S. K.,** Anti-proteolytic activity of some triterpenoids, *Int. J. Crude Drug Res.,* 21, 93, 1983.

53. **Chatterjee, K. K.,** Treatment of cancer: a prelude, *Indian Med. Record (Calcutta),* 81, 101, 1961.

54. **Pettit, G. R., Barton, D. H. R., Herald, G. L., Polonsky, J., Schmidt, J. M., and Connolly, J. D.,** Evaluation of limonoids against the murine P388 lymphocytic leukemia cell line, *J. Nat. Prod.,* 46, 379, 1983.

55. **Sadagopan, V. R., Johri, T. S., and Reddy, V. R.,** Feeding value of neem seed meal in broiler and layer diet (chickens), *Indian Poultry Gazz.,* 65, 136, 1981.

56. **Sadagopan, V. R., Johri, T. S., Reddy, V. R. and Panda, B. K.,** Feeding value of neem seed meal for starter chicks, *Indian Vet. J.,* 59, 462, 1982.

57. **Subbarayudum, D. and Reddy, V. R.,** in *Proc. 4th Poultry Sci. Symp.,* Bhuvaneshwar, India, 1975.

58. **Thakur, B. S., Deb, S., Ladukar, O. N., and Deb, R. N.,** Utilization of neem cake in broiler mash, *Indian Poultry Rev.,* 8(21), 13, 1977.

59. **Bhandari, D. S. and Joshi, M. S.,** The effect of feeding deoiled neem cake on health of sheep, *Indian Vet. J.,* 51, 659, 1974.

60. **Gupta, R. S. and Bhaid, M. U.,** Studies on agri-industrial by-products (deoiled neem fruit cake in sheep feed composition), *Indian Vet. J.,* 58, 311, 1981.

61. **Vijjan, V. K., Tripathy, H. C., and Parihar, N. S.,** A note on the toxicity of neem (*Azadirachta indica*) seed cake in sheep, *J. Environ. Biol.,* 3, 47, 1982.

62. **Singh, N. P. and Patnayak, B. C.,** A note on the palatability of neem tree leaves for sheep, *Food Farm. Agric.,* 14, 47, 1981.

63. **Perera, H. C.,** Indigenous drugs in the treatment of livestock, *Indian Vet. J.,* 17, 261, 1941.

64. **Christopher, J.,** Neem-seed cake is also good for cattle, *J. Indian Farm.,* 20(1), 38, 1970.

65. **Ludri, R. S. and Arora, S. F.,** Neem seed meal utilization for growth and its influence on ^{35}S sulfur isotope incorporation into microbial protein, buffalo calves, *J. Nucl. Agric. Biol.,* 5, 67, 1976.

66. **Rao, B. S.,** Alkali-Treated Neem (*Azadirachta indica*) Seed Cake as a Cattle Feed, B.S. thesis, Rohilkhand University, Bareilly, India, 1977.

67. **Nath, K. A., Vijjan, V. K., and Ranjhan, S. K.,** Alkali-treated neem seed cake as a livestock feed, *J. Agric. Sci. Cambridge,* 90, 531, 1978.

68. **Devasia, P. A.,** Studies on the feeding value of neem (*Azadirachta indica* Juss) seed cake

for cattle, *Kerala J. Vet. Sci.*, 10, 182, 1979.

69. **Singh Bedi, P. S., Vijjan, V. K., and Ranjhan, S. K.,** Utilization of neem (*Azadirachta indica*) seed cake and its influence on nutrient digestibilities in buffaloes, *Indian J. Dairy Sci.*, 28, 104, 1975.

70. **Singh Bedi, P. S., Vijjan, V. K., and Ranjhan, S. K.,** Effect of neem (*Azadirachta indica*) on growth and digestibilities of nutrients in cross-bred calves, *Indian J. Anim. Sci.*, 65, 618, 1977.

71. **Arora, S. P., Singhal, K. K., and Ludri, R. S.,** Nutritive value of neem seed cake (*Melia indica*). Cattle, *Indian Vet. J.*, 52, 867, 1975.

72. **Pyne, A. K., Moitra, D. N., and Gangopadhyay, P.,** Studies on the composition of milk with the use of neem seed expeller cake (*Azadirachta indica*) on lactating buffaloes, *Indian Vet. J.*, 56, 223, 1979.

73. **Gangopadhyay, P., Maitra, D. N., and Pyne, A. K.,** Studies on the blood constituents with the use of neem seed expeller cake (*Azadirachta indica*) in lactating Murrah buffaloes, *Indian Vet. J.*, 56, 979, 1979.

74. **Maitra, D. N., Roy, S., and Duttagupta, R.,** Efficiency of utilization of digestible and metabolisable energy for milk production with neem cake, *Indian J. Dairy Sci.*, 35, 368, 1982.

75. **Flinders, C.,** Laurel's kitchen, *The Washington Post,* Washington, D.C., January 26, 1983.

76. **Patrao, M. R.,** Rare neem tree with sweet leaves, *Neem Newsl.*, 2(3), 34, 1985.

77. **Fujiwara, T., Takeda, T., Ogihara, Y., Shimizu, M., Nomura, T., and Tomita, Y.,** Studies on the structure of polysaccharides from the bark of *Melia azadirachta, Chem. Pharm. Bull.*, 30, 4025, 1982.

78. **Fujiwara, T., Sugishita, E., Takeda, T., Ogihara, Y., Shimizu, M., Nomura, T., and Tomita, Y.,** Further studies on the structure of polysaccharides from the bark of *Melia azadirachta, Chem. Pharm. Bull.*, 32, 1385, 1984.

79. **Anderson, D. M. W., Bell, P. C., Gill, M. C. L., McDougall, F. J., and McNab, C. G. A.,** The gum exudates from *Chloroxylon swietenia, Sclerocarya caffra, Azadirachta indica,* and *Moringa oleifera, Phytochemistry,* 25, 247, 1986.

80. **Kahn, M. W., Khan, A. M., and Saxena, S. R.,** Rhizosphere fungi and nematodes of eggplant as influenced by oilcake amendments, *Indian Phytopathol.*, 27, 480, 1974.

81. **Kahn, M. W., Alam, M. M., Khan, A. M., and Saxena, S. K.,** Effect of water-soluble fractions of oil-cakes and bitter principles of neem on some fungi and nematodes, *Acta Bot. Indica,* 2, 120, 1974.

82. **Ahmed, S. R. and Agrihotri, J. P.,** Antifungal activity of some plant extracts, *Indian J. Mycol. Plant Pathol.*, 7, 180, 1977.

83. **Aguda, R. M. and Rombach, M. C.,** Effect of neem oil on germination and sporulation of the entomogenous fungus *Metarhizium anisopliae, Int. Rice Res. Newslett.*, 11(4), 34, 1986.

84. **Kahn, M. and Wassilew, S. W.,** The effect of raw material of the neem tree, neem oils, and neem extracts on fungi pathogenic to humans, Abstr. 3rd Int. Neem Conf., Nairobi, Kenya, July 1986, 65.

85. **Alice, D. and Rao, A. V.,** Antifungal effects of plant extracts on *Drechslera oryzae* in rice, *Int. Rice Res. Newslett.*, 12, 28, 1987.

86. **Watanabe, I., Subhudi, B. P. R., and Aziz, T.,** Effect of neem cake on the population and nitrogen fixing activity of blue-green algae in flooded soil, *Curr. Sci.*, 50, 937, 1981.

87. **McGregor, J.,** personal communication, 1979.

88. **Klocke, J. A.,** personal communication, 1985.

89. **Jeter, W. S.,** personal communication, 1980.

90. **Jacobson, M.,** Neem research in the U.S. Department of Agriculture: chemical, biological and cultural aspects, in *Natural Pesticides from the Neem Tree (Azadirachta indica A. Juss),* Schmutterer, H., Ascher, K. R. S., and Rembold, H., Eds., GTZ Press, Eschborn, West Germany, 1981, 33.

91. **Rojanapo, W., Suwanno, S., Somjaree, R., Glinsukwu, T., and Thebtaranoni, Y.,** Mutagenic and antibacterial activity testing of nimbolide and nimbic acid, *J. Sci. Soc. Thailand,* 11, 177, 1985.

92. **Uwaifo, A. O.,** The mutagenicities of seven coumarin derivatives and a furan derivative (nimbolide) isolated from three medicinal plants, *J. Toxicol. Environ. Health,* 13, 521, 1984.

93. **Jongen, W. M. F. and Koeman, J. H.,** Mutagenicity testing of two tropical plant materials with pesticide potential in *Salmonella typhimurium. Phytolacca dodecandra* berries and oil from seeds of *Azadirachta indica, Environ. Mutagenesis,* 5, 687, 1983.

94. **Schneider, B. H.,** The effect of neem leaf extracts on *Epilachna varivestis* and *Staphylococcus aureus,* Abstr. 3rd Int. Neem Conf., Nairobi, Kenya, July 1986, 73.

95. **Singh, R.,** Inactivation of potato X-virus by plant extracts, *Phytopathol., Mediterr.,* 10, 211, 1971.

96. **Verma, V. S.,** Chemical compounds from *Azadirachta indica* as inhibitors of potato X-virus, *Acta Microbiol.,* 68, 9, 1974.

97. **Tella, A.,** The effects of *Azadirachta indica* in acute *Plasmodium berghei* malaria, *W. Afr. J. Pharmacol. Drug Res.,* 3, 80P, 1976.

98. **Heyneman, D.,** personal communication,

1981.

99. **Stone, R. J.,** personal communication, 1986.

100. **Muley, E. V.,** Biological and chemical control of the vector snail *Melania scabra* (Gastropoda:Prosobranchia), *Bull. Zool. Survey India,* 1, 1, 1978.

101. **Reynaud, P. A.,** Control of the azolla pest *Limnea natalensis* with molluscicides of plant origin, *Int. Rice Res. Newslett.,* 11(3), 27, 1986.

102. **Mikolajczak, K. L. and Reed, D. K.,** Extractives of seeds of the Meliaceae: effects on *Spodoptera frugiperda* (J. E. Smith), *Acalymma vittatum* (F.), and *Artemia salina* Leach, *J. Chem. Ecol.,* 13, 99, 1987.

103. **Anon.,** Neem tree may be source of safe insecticides, *Int. Rice Res. Inst. Reptr.,* 2, 2, 1982.

104. **Grant, I. F.,** Effects of aqueous neem kernel extracts on ostracod (class Crustacea) development and population density in lowland rice fields, Abstr. 3rd Int. Neem Conf., Nairobi, Kenya, July 1986, 63.

105. **Roessner, J. and Zebitz, C. P. W.,** The effect of soil treatment with neem products on earthworms (Lumbrididae), Abstr. 3rd Int. Neem Conf., Nairobi, Kenya, July 1986, 72.

106. **Jain, H. K.,** Neem in agriculture, *Indian Agric. Res. Inst. Bull.,* No. 40, 54, 1983.

107. **Vijayalakshmi, K., Gaur, H. S., and Goswami, B. K.,** Neem for the control of plant parasitic nematodes, *Neem Newslett.,* 2(4), 35, 1985.

108. **Redknap, R. S.,** The use of crushed neem berries in the control of some insect pests in Gambia, in *Natural Pesticides from the Neem Tree (Azadirachta indica A. Juss),* Schmutterer, H., Ascher, K. R. S., and Rembold, H., Eds., GTZ Press, Eschborn, West Germany, 1981, 205.

109. **Ravindran Nair, K. K., Kamalakshmi Amma, P. L., and Kuriyan, K. J.,** Use of neem cakes for the control of root knot nematodes in brinjal, Abstr. Natl. Symp. on Soil Pests and Soil Organisms, Banaras Hindu University, Varanasi, India, October 1984, 72.

110. **Anon.,** Nematicidal and egg hatching inhibitor, *Indian Agric. Res. Inst. Newslett.,* 4, 3, 1982.

111. **Saxena, S. K. and Khan, A. M.,** Different aspects of the control of nematodes by neem, Abstr. 2nd Int. Neem Conf., Rauisch-Holzhausen, West Germany, May 1983, 43.

112. **Siddiqui, A., Alam, M. M., and Saxena, S. K.,** Studies on the nema-toxicity of neem and bakain, Abstr. 2nd Int. Neem Conf., Rauisch-Holzhausen, West Germany, 1983, 51.

113. **Jacobson, M., Stokes, J. B., Warthen, J. D., Jr., Redfern, R. E., Reed, D. K., and Webb, R. E.,** Neem research in the U.S. Department of Agriculture: an update, Abstr. 2nd Int. Neem Conf., Rauisch-Holzhausen, West Germany, May 1983, 29.

114. **Husain, S. I. and Masood, A.,** Effect of some plant extracts on larval hatching of *Meloidogyne incognita* (Kofoid & White) Chitwood, *Acta Bot. Indica,* 3, 142, 1975.

115. **Siddiqui, M. A. and Alam, M. M.,** Further studies on the nematode toxicity of margosa and Persian lilac, *Neem Newslett.,* 2(4), 43, 1985.

116. **Johnson, D. T.,** personal communication, 1981.

117. **Devakumar, C., Goswami, B. K., and Mukerjee, S. K.,** Nematicidal principles from neem (*Azadirachta indica* A. Juss). I. Screening of neem kernel fractions against *Meloidogyne incognita* (Kofoid & White) Chitwood, *Neem Newslett.,* 2(2), 21, 1985.

118. **Sharma, H. L., Vimal, O. P., and Prasad, D.,** Nematicidal and manurial value of some oil cakes, *Int. J. Trop. Plant Dis.,* 3, 51, 1985.

119. **Bhattacharya, D.,** Studies on Efficacy of Groundnut and Neem Oil Cakes in Comparison to a Systemic Nematicide, Aldicarb, Against Root Knot Nematode, *Meloidogyne incognita,* on Tomato, M.S. thesis, Indian Agricultural Research Institute, New Delhi, 1984.

120. **Venkata Rao, C., Mani, A., and Kameswara Rao, P.,** Effect of plant products on egg hatch and larval mortality of *Meloidogyne incognita, Proc. Ind. Acad. Sci. (Anim. Sci.),* 95, 397, 1986.

121. **Roessner, J. and Zebitz, C. P. W.,** Effect of neem products on nematodes and growth of tomato (*Lycopersicon esculentum*) plants, Abstr. 3rd Int. Neem Conf., Nairobi, Kenya, July 1986, 64.

122. **Sitaramaiah, K.,** Neem oil cake amendment and its effect on plant-parasitic nematodes, Abstr. 2nd Int. Neem Conf., Rauisch-Holzhausen, West Germany, May 1983, 53.

123. **Siebeneicher, F.,** Investigations on the Use of Neem (*Azadirachta indica* A. Juss) to Control Gall-forming Nematodes (*Meloidogyne* spp.) in Togo, Ph.D. thesis, University of Giessen, West Germany, 1982.

124. **Dasgupta, D. R. and Ganguly, A. K.,** Control of plant parasitic nematodes by margosa, Abstr. Natl. Semin. on Neem in Agriculture, Indian Agricultural Research Institute, New Delhi, April 1983, 6.

125. **Vijayalakshmi, K. and Prasad, S. K.,** Effect of some nematicides, oilseed cakes and inorganic fertilizers on nematodes and crop growth, *Ann. Agric. Res.,* 3, 133, 1982.

126. **Mishra, S. D. and Gaur, H. S.,** Control of nematodes infesting mung with nematicidal seed treatment and field applications of dasanit, aldicarb and neem cake, Abstr. Natl. Symp. on Soil Tests and Soil Organisms,

Banaras Hindu University, Varanasi, India, October 1984, 71.

127. **Saxena, P. K., Chabra, H. K., and Kiran, J.,** *Effect of certain soil amendments and nematicides on the population of nematodes infesting grape-vines, Z. Angew. Zool.,* 64, 325, 1977.

128. **Siddiqui, M. A. and Alam, M. M.,** Evaluation of nematocidal properties of different parts of margosa and Persian lilac, *Neem Newslett.,* 2(1), 1, 1985.

129. **Husain, S. I. and Masood, A.,** Nematicidal action of plant extracts on parasitic nematodes, *Geobios,* 2, 74, 1975.

130. **Egunjobi, O. A. and Afolami, S. O.,** Effects of water-soluble extracts of neem (*Azadirachta indica*) on *Pratylenchus brachyurus* and on maize, *J. Nematol.,* 7, 321, 1975.

131. **Egunjobi, O. A. and Afolami, S. O.,** Effects of neem (*Azadirachta indica*) leaf extracts on populations of *Pratylenchus brachyurus* and on the growth and yield of maize, *Nematologica,* 22, 125, 1976.

132. **Gill, J. S. and Lewis, C. T.,** Systemic action of an insect feeding deterrent, *Nature (London),* 232, 402, 1971.

133. **Kaplan, D. T.,** personal communication, 1984.

134. **Yassin, A. M., Loof, D. A. A., and Oostenbrink, J.,** Plant parasitic nematodes in the Sudan, *Nematologica,* 16, 567, 1968.

135. **Yassin, A. M.,** A note on *Longidorus* and *Xiphinema* species from the Sudan, *Nematol. Mediterr.,* 2, 141, 1974.

136. **El Amin, E. T. M., Ahmed, M. A., and Mirghani, A. A.,** An Introductory Note to *Azadirachta indica* A. Juss in the Sudan, Gezira Research Station, Wad Medani, Sudan, 1983.

137. **Attri, B. S. and Ravi Prasad, G.,** Neem oil extractive — an effective mosquito larvicide, *Indian J. Entomol.,* 42, 371, 1980.

138. **Ketkar, C. M.,** personal communication, 1979.

139. **El Amin, E. T. M., Ahmed, M. A., and Mirghani, A. A.,** An Introductory Note to *Azadirachta indica* A. Juss in the Sudan, unpublished report, 1983.

140. **Doria, J. J.,** Neem: the tree insects hate, *Garden,* July-August, 1981, 8.

141. **Kubo, I. and Nakanishi, K.,** Some terpenoid insect antifeedants from tropical plants, in *Advances in Pesticide Science,* IUPAC, Part 2, Geissbuhler, H., Ed., Pergamon Press, Elmsford, N.Y., 1979, 284.

142. **Sengupta, S.,** *World Agrochem. News,* No. 8, 12, 1986.

143. **Singh, Y. P., Bahga, H. S., and Vijjan, V. K.,** Toxicity of water extract of neem (*Azadirachta indica* A. Juss) berries in poultry birds, *Neem Newslett.,* 2(2), 17, 1985.

144. **Starr, R. I., Besser, J. F., and Brunton, R. H.,** A laboratory method for evaluating chemicals as bird repellents, *J. Agric. Food*

Chem., 12, 342, 1964.

145. **Schafer, E. W., Jr. and Jacobson, M.,** Repellency and toxicity of 55 insect repellents to red-winged blackbirds (*Agelaius phoeniceus*), *J. Environ. Sci. Health,* A18, 493, 1983.

146. **Asplin, F. D. and Carnoghan, R. B. A.,** *Vet. Rec.,* 73, 1215, 1961.

147. **Sinniah, D., Baskaran, G., and Leong, K. L.,** Post mortem studies in mice following experimentally induced margosa oil poisoning and some characteristics of the toxin(s), Abstr. 15th Malaysian-Singapore Cong. of Medicine, Kuala Lumpur, Malaysia, December 1980, 88.

148. **Vijjan, V. K. and Tandan, S. K.,** Drug-interaction feasibility of water extract of neem (*Azadirachta indica*) seed cake, *Neem Newslett.,* 2(2), 15, 1985.

149. **Bhide, N. K., Mehta, D. J., Altekar, W. W., and Lewis, R. A.,** Toxicity of sodium nimbidinate, *Indian J. Med. Sci.,* 12, 146, 1958.

150. **Pillai, N. R. and Santhakumari, G.,** Toxicity studies on nimbidin, a potential antiulcer drug, *Planta Med.,* 50, 146, 1984.

151. **Qadri, S. S. H., Usha, G., and Jabeen, K.,** Subacute dermal toxicity of neemrich-100 (tech.) to rats, *Int. Pest Control,* 26, 18, 1984.

152. **Sardeshpande, P. D.,** Carcinogenic Potency of Neem Seed Cake in Rat, Department of Pathology, Bombay Veterinary College, India, 1976.

153. **Ali, B. H.,** The toxicity of *Azadirachta indica* leaves in goats and guinea pigs, *Vet. Human Toxicol.,* 29, 16, 1987.

154. **Ali, B. H. and Salih, A. M. M.,** Suspected *Azadirachta indica* toxicity in sheep, *Vet. Rec.,* 111, 494, 1982.

155. **Chopra, R. N., Badhwar, R. L., and Ghosh, S.,** *Poisonous Plants of India,* Vol. 1 (revised), Indian Council of Agricultural Research, New Delhi, 1965, 245.

156. **Bhide, N. K., Shah, N. J., Sardesai, H. V., and Sheth, U. K.,** Clinical Studies With Sodium Nimbidinate, unpublished data.

157. **Sinniah, D. and Baskaran, G.,** Margosa oil poisoning as a cause of Reye's syndrome, *Lancet,* 487, 1981.

158. **Sinniah, D., Baskaran, G., Looi, L. M., and Leong, K. L.,** Reye-like syndrome due to margosa oil poisoning: report of a case with postmortem findings, *Am. J. Gastroenterol.,* 77, 158, 1982.

159. **Sinniah, D., Varghese, G., Baskaran, G., and Koo, S. H.,** Fungal flora of neem (*Azadirachta indica*) seeds and neem oil toxicity, *Malays. Appl. Biol.,* 12, 1, 1983.

160. **Jacobson, M.,** Pharmacological and toxicological effects of neem and chinaberry on warm-blooded animals, *Neem Newslett.,* 3(4), 39, 1986.

161. **Sharma, V. N. and Saksena, K. P.,** Sper-

micidal action of sodium nimbidinate, *Indian J. Med. Res.*, 47, 322, 1959.

162. **Sharma, V. N. and Saksena, K. P.,** Sodium nimbidinate — in vitro study of its spermicidal action, *Indian J. Med. Sci.*, 13, 1038, 1959.

163. **Baker, J. R., Ransom, R. M., and Tynen, J.,** The spermicidal powers of chemical contraceptives, *J. Hyg.*, 37, 474, 1937.

164. **Choudhary, R. R.,** Plants With Possible Antifertility Activity, Special Report, Indian Council of Medical Research, New Delhi, 1966, 3.

165. **Deshpande, V. Y., Mandulkar, K. N., and Sadre, N. L.,** Male antifertility activity of *Azadirachta indica* in mice. A preliminary report, *J. Postgrad. Res.*, 26, 167, 1980.

166. **Deshpande, V. Y., Mandulkar, K. N., and Sadre, N. L.,** Male antifertility activity of *Azadirachta indica* in mice. Further study, 4th Asian Symp. on Medicinal Plants and Spices, Bangkok, Thailand, 1981.

167. **Sadre, N. L., Deshpande, V. Y., Mandulkar, K. N., and Nandal, D. H.,** Male antifertility activity of *Azadirachta indica* in different species, in *Natural Pesticides from the Neem Tree and Other Tropical Plants,* Schmutterer, H. and Ascher, K. R. S., Eds., GTZ Press, Eschborn, West Germany, 1984, 473.

168. **Khare, A. K., Srivastava, M. C., Sharma, M. K., and Tewari, J. P.,** Antifertility activity of neem oil in rabbits and rats, *Probe, 23,* 90, 1984.

169. **Krause, W. and Adami, M.,** Extracts of neem (*Azadirachta indica*) seed kernels do not inhibit spermatogenesis in the rat, in *Natural Pesticides from the Neem Tree and Other Tropical Plants,* Schmutterer, H. and Ascher, K. R. S., Eds., GTZ Press, Eschborn, West Germany, 1984, 483.

170. **Sinha, K. C., Riar, S. S., Tiwary, R. S., Dhawan, A. K., Bardhan, J., Thomas, P., Kain, A. K., and Jain, R. K.,** Neem oil as a vaginal contraceptive, *Indian J. Med. Res.,* 79, 131, 1984.

171. **Sinha, K. C., Riar, S. S., Bardhan, J., Thomas, P., Kain, A. K., and Jain, R. K.,** Anti-implantation effect of neem oil, *Indian J. Med. Res.,* 80, 708, 1984.

172. **Riar, S. S.,** personal communication, 1985.

173. **Lal, R., Sankarayanarayana, A., Mathur, V. S., and Sharma, P. L.,** Antifertility effect of neem oil in female albino rats by the intravaginal and oral routes, *Indian J. Med. Res.,* 83, 89, 1986.

174. **Tewari, R. K., Mathur, R., and Prakash, A. O.,** Post-coital antifertility effect of neem oil in female albino rats, *IRCS Med. Sci.,* 14, 1005, 1986.

175. **Pathak, P. H. and Krishna, S. S.,** Neem seed oil, a capable ingredient to check rice moth reproduction (Lepid., Galleriidae), *Z. Angew. Zool.,* 100, 33, 1985.

176. **Von der Heyde, J., Saxena, R. C., and Schmutterer, H.,** Effects of neem derivatives on growth and fecundity of the rice pest *Nephotettix virescens* (Homoptera:Cicadellidae), *Z. Pflanzenkr. Pflanzenschutz,* 92, 346, 1985.

177. **Schmutterer, H.,** Fecundity-reducing and sterilizing effects of neem seed kernel extracts in the Colorado potato beetle, *Leptinotarsa decemlineata* Say, Abstr. 3rd Int. Neem Conf., Nairobi, Kenya, July 1986, 40.

178. **Schmidt, G. H.,** Studies of the sterilizing effect of neem oil in ants, Abstr. 3rd Int. Neem Conf., Nairobi, Kenya, July 1986, 41.

179. **Anon.,** Sensal, *Neem Newslett.,* 3(3), 37, 1986.

9 The Commercialization of Neem

Robert O. Larson
Vikwood Botanicals, Inc.
Sheboygan, Wisconsin

TABLE OF CONTENTS

I. HISTORICAL PERSPECTIVE

The history of commerce in the uses of parts of the neem tree is shrouded in the mystery and lore of the Vedic period of India (which began about 4000 years ago) when the many uses of neem tree parts were developed into Ayurvedic medicine (forms of treatment developed to utilize natural herbs, seeds, leaves, flowers, etc. both in home remedies as well as in primitive pharmaceutical treatments by a caste of nomadic mendicants). Sanskrit writings dating back to this period refer to curatives for internal disorders such as diabetes, stomach aches, malarial fever, stomach worms, etc. These were derived primarily from offerings of the tree. To this day, this caste of traveling physicians meets annually in East Central India to exchange herbs, ideas, tree parts, and applicative details as to their uses, performing ancient rites at a shrine in an old neem tree, the altar being part of the tree itself. These mendicants depart shortly thereafter touring their territory and bringing cures, hope, a touch of the divine and a large dose of solace in exchange for a few rupees or bartered goods. This was the beginning of commerce in neem.

In the distant past, aside from home use (seeds and leaves used as repellants) to protect food in bags, boxes, pots, etc. the "commercial" development of neem arose from the expression of oil from the seed kernels to fill the demand for lamp fuel, medicinal body lotions and purgatives, as well as soap bases. The current "stand" of neem trees in India is estimated at 14,000,000 trees, although this figure is probably outdated and is currently much larger. It is estimated that less than 20% of the seed crop is harvested due to the scattered nature of the trees, and the widely separated oil mills.[1] The seed is usually taken to the mill with the fruit on, which soon becomes darkened through oxidation and becomes hardened to the seed hull. The oil yield is relatively low due to the lack of sophisticated equipment and procedures, thus rendering seed collection and oil expression seasonal and relatively ineffective.

To date there appear to be no large plantations developed in India which might encourage capable chemical processors to develop better extractive systems, and develop efficient systems to produce stabilized neem seed extracts as inexpensive and readily available pesticides at the corner market. It has been estimated[2] that neem applications to repel insects could be as low as one tenth the cost of malathion, which has recently been used even in the poorest countries of Africa due to the lack of knowledge about the use of the neem seed available in their immediate areas.

II. DISCOVERY AND USE OF CRUDE NEEM KERNEL SUSPENSIONS

The application of a water extract as an insect repellant to crop pests is fairly recent, and early inquiries into the methods of application and the effective constituents from the seed, leaf, and leaf cake are largely Indian contributions. Dried and ground neem seed powder mixed with water even at a ratio as low as 100 g of seed to 10 l of water makes an effective insect repellant to a wide variety of food consuming pests when filtered and used as a simple foliar spray.[3] Only in the 1900s was it discovered that a crude neem extract could act as a repellant, an antifeedant, and more recently, as a growth regulator due to hormonal disruption. These discoveries were refined into new areas of inquiry in both India and Africa against both tobacco and coffee bean pests. The dispersion of this information is largely the result of migrations of Indians to Africa in the 1800s and 1900s. Neem seeds were taken by the family to grow a neem tree at a new home site in a new country to ensure a supply of home remedies originally based on Ayurvedic medicines of the distant past.

Prior to this, the neem tree was translocated by birds and fruit bats carrying off the fruit for consumption at a more convenient spot, discarding the seed at some distance from the source. However, the varied uses of the active ingredients of the neem tree lagged far behind the organized plantings, and some of the central and west African countries, where the tree is largely ignored, have yet to learn of the many uses of the neem tree parts. European colonists set out the trees along highways in some countries at measured intervals of about 10 m, and there is no evidence that much effective use is made of the seed for any reason and no industry appears to be set up even for expressing the neem oil. The plantings that do exist were apparently made for soil retention, shade, and as windbreaks.

Recent efforts via three International Neem Tree Conferences (two in Germany and one in Nairobi, Kenya) plus the efforts of the GTZ (German Association for Technical Cooperation) and various altruistically oriented agro-groups have brought simple but easily obtainable crop protection methods to the millions of small villages in affected areas of Africa. Elsewhere, ill-planned applications of commercial pesticides such as malathion were dispersed by airplanes with relatively expensive and, in some cases, futile results bringing to question the high costs and overall legacy of using questionable pesticides.

Led largely by Indian efforts, various solvent extracts from neem seeds and leaves were developed and tried on a wide variety of edible crops, as well as tobacco, etc. where neem leaf extracts were found to possess antiviral activity when administered to young tobacco seedlings under 20 d of age.[4] Crude neem seed extracts repelled feeding by a host of migratory insects which on occasion preferred to starve rather than to feed on the sprayed leaves.

By the late 1940s and 1950s, serious inquiry was being undertaken in many Indian universities and institutes with the objective being the development of a simple extract with quantified constituents in terms of insect control activity. A wide spectrum of solvents and solvent combinations was tried, usually defined in terms of "percentages of neem kernel extract per liters of water", for example. The failure to define the active ingredient or ingredients in the extract resulted in mixed and contradictory tests.

In the 1960s highly technical analyses were attempted in England and elsewhere which resulted in the first structuring of the mysterious active ingredient azadirachtin (AZ), a tetranortriterpenoid mainly responsible for the unique insecticidal qualities of the extract. The first structure for azadirachtin ($C_{35}H_{44}O_{16}$) postulated by Butterworth and Morgan[5] has been revised more recently.[6,7] Other significant research and development has been investigated by the group of scientists at the U.S. Department of Agriculture (USDA), Beltsville, MD, under the direction of Martin Jacobson, as well as by several groups in Germany.

Through empirical testing throughout the world, azadirachtin-rich extracts of known and predictable activity were developed despite the fact that the quantums of the various ingredients were ill-defined. Efforts to prepare and store large quantities of relatively pure azadirachtin were not possible due to high cost and rapid decomposition of the compound when exposed to air, heat, moisture, and ultraviolet light. Even today a pure azadirachtin standard must be stored under nitrogen in a desiccator and kept at $-40°C$ to hold its purity. Despite this, azadirachtin (of unknown purity) is presently being offered for sale by a German concern at a cost of $5.00/mg.[8] Formulations varied with unknown quantities of azadirachtin being extracted and incorporated into the trial applications, making exchange of information difficult. Variability of the efficacy of the formulations gave puzzling responses and investigations were begun on the effect of tree sitings, cultivars, quality of seed, soil types, and variables in the insects tested.

The solution to the early problems appeared to be selection of fresh neem seeds to give a predictable extract with definable storability, or moving into the area of pure chemistry by synthesizing azadirachtin with known activity, thereby eliminating the uncertainty of working with a volatile botanical extract. Such synthesis in the near future does not appear to be promising in view of the complexity of the molecule.

III. USDA-BELTSVILLE, MD, U.S.A. INQUIRY

During the mid and later 1970s, the USDA at Beltsville, MD began efforts to test topically and in diets the efficacy of solutions containing up to 90% pure azadirachtin extracted from neem seeds, obtained in India, on a variety of insects. By pure chance, Vikwood Ltd. was brought into the program as a result of being able to locate seeds of high azadirachtin content in India. This resulted in a collaborative agreement whereby information generated by USDA and Vikwood Ltd. would be shared in a common effort to develop an effective and predictable shelf-stable extract for the commercial market.

I was personally intrigued by the possibility of a program targeted to remove the use of toxic pesticides from American horticulture and agriculture, since the horror stories of indiscriminate use of toxic pesticides was just beginning. My recollection of witnessing the repellant effects of the neem tree in Nagpur, India in 1973 rushed back with memories of how my hosts showed that food grains could be completely protected and stored in the home by throwing a handful of defruited fresh neem seeds into a jar or container holding their food grains. Additionally, neem leaves were placed in books to repel silverfish, inserted in wool clothing to discourage egg laying by moths, and placed under mattresses to repel crawling night insects. The neem flower broken into bits, placed on rice, and served with three kinds of oil was used as an effective stomachic for children. The ubiquitous neem twig served, and still does, over half a billion people a day as an effective toothbrush with clear evidence of a sharp reduction in cavities and gum diseases. Only the wealthy and western-oriented people of India have abandoned this practice, somewhat to the detriment of their health. It seems every Indian woman knows when and how to use the neem leaves in a thin paste, after being ground and boiled, and rubbed on the skin as an unguent to cure scabies and the external effects of chicken pox and smallpox. This appears to be common knowledge everywhere in India.

IV. DEVELOPMENT OF MARGOSAN-O

Thus began an 8-year effort to develop, with phytochemists in the Wisconsin area, a stable but inexpensive neem extract with predictable activity and good shelf stability. We were encouraged by early efforts to locate seeds of even greater azadirachtin content and sought out seed from other countries in tropical areas. After a year or so of trials using a variety of solvent combinations, we settled on hexane as the most effective solvent to remove the oil from the seed which otherwise impeded the extraction of azadirachtin with ethanol. Virtually all tests showed that complete removal of the seed oil via hexane, or other hydrocarbon solvents, gave a mixture from which an azadirachtin-rich extract could be made with no loss of the active component.

Ultimately the system employed gave extract yields of 5400 ppm AZ or better. Regretfully, our experience paralleled those of the USDA with only short-term shelf stability being achieved. A revision of the formula to an arbitrary figure of 3000 ppm AZ gave us reasonable stability but only when controlled by adjusting the pH of the extract and the addition of a number of emulsifiers and stabilizers. The first product showed very little loss in activity over a 9-month period and was then subjected to the rigorous toxicity studies required by the U.S. Environmental Protection Agency (EPA) to qualify for registration. Due to the "uniqueness" of the product, new tests were "tailor-made" for the testing of the formulation, designated "Margosan-O", since the traditional protocols for testing a product of this type would not show the effect of azadirachtin on the various birds, animals, and fish selected for these tests.

With the USDA-Beltsville and other public and private institutions in the U.S. and abroad doing efficacy studies with the product, Vikwood Ltd. forged ahead with the toxicity studies required and submitted the final data for review in 1983. Despite earlier promises for expeditious review of this product, due to the obvious need to replace toxic pesticides with a product safe and natural such as Margosan-O, the EPA review took over $1\frac{1}{2}$ years but finally culminated with the registration of MARGOSAN-O;® in July 1985, EPA Reg. No. 58370-1.

V. TOXICITY STUDIES REQUIRED BY THE EPA AND THEIR RESULTS

Test 1, Avian Single-Dose Oral LD_{50}. Margosan-O was administered to mallard ducks in order to determine a dose lethal to 50% of the duck population. Dose levels of Margosan-O ranged at 1 to 16 ml/kg of body weight. Observations showed no negative effects and all ducks remained active and healthy throughout the 14-d experimental period. The acute LD_{50} of Margosan-O to mallard ducks is in excess of 16.0 ml/kg.

Test 2, Avian Dietary LC_{50} (lethal concentration) with Bobwhite Quail. The birds were given their basal diet, with additions of Margosan-O ranging from 1000 to 7000 ppm. Observations showed no negative effects and the quail were active and healthy throughout the 5-d test period and 3-d recovery phase. The acute oral LC_{50} of the Margosan-O to bobwhite quail is therefore in excess of 7000 ppm.

Test 3, Avian Dietary LC_{50} Study with Mallard Ducks. The ducks were given a basal diet, plus Margosan-O ranging from 1000 to 7000 ppm concentrate for 5 d. No mortalities resulted. The ducks were active and healthy throughout the test and recovery phases. The acute LC_{50} of the test material to mallard ducks is therefore in excess of 7000 ppm.

Test 4 (No. 1), Acute toxicity of Margosan-O to Rainbow Trout. This test involved a 96-h LC_{50} of Margosan-O at various concentrations. The LC_{50} was 8.8 ml of Margosan-O per liter of water, and the 96-h no-observed-effect concentration was 5 mg/l.

Test 4 (No. 2), Acute Toxicity of Margosan-O to Bluegill Sunfish. The results showed a 96-h LC_{50} of 37 mg/l and a no-observed-effect 96-h level of 20 mg/l. In the fish bioassay tests with trout and sunfish there were behavioral responses in static water which could probably not occur in moving water.

Test 5, Acute Toxicity of Margosan-O to *Daphnia magna*, the Water Flea. The test was done on newly molted instars less than 20 h old which were placed in a fresh aquatic habitat for up to 48 h. The LC_{50} was 13 mg/l and the no-observed-effect concentration at 48 h was less than 10 mg/l. Therefore, the toxicity value obtained was well within the expected range, but it indicates that Margosan-O will affect primitive aquatic invertebrates under static conditions.

Test 6, Acute Oral Toxicity. Rats were dosed once and then observed for 14 d for abnormal behavior or mortality. No negative effects were observed and the acute oral toxicity of the test material was in excess of 5 ml/kg, the limit of the required test.

Test 7 (No. 1), Acute Dermal Toxicity. A nonpermeable patch containing 2 ml/kg body weight of Margosan-O was placed over small shaved areas on a group of albino rabbits. No mortality resulted and the acute dermal toxicity (LC_{50}) of Margosan-O was in excess of 2 ml/kg.

Test 7 (No. 2), Primary Skin Irritation. Albino rabbits were treated with Margosan-O applied under patches on shaved areas and on abraded areas. The results showed low to moderate primary irritation to the shaved area patch and high to moderate irritation to the abraded area.

Test 8, Acute Inhalation Study. Albino rats were exposed to a total of 15.8 g of test material (estimated concentration of 43.9 mg/l/h) for 4 h. (This test was recently repeated and reported in terminology more acceptable to the EPA). The LC_{50} for the Margosan-O in the inhalation test was in excess of 43.9 mg/l/h, the limit of the test.

Test 9, Modified Eye Irritation. Margosan-O was administered to one washed and one unwashed eye of albino rabbits. Over 7 d both eyes showed minimal irritation.

Test 10, Immune Response. The effect of Margosan-O on the hematology and serum electrophoretic pattern of rats, strain Sprague-Dawley, was determined. Eight male and eight female rats, each weighing between 200 and 250 g, were anesthetized by means of CO_2 and weighed. A 3-ml sample of blood was taken via cardiac puncture and the blood studied. Five male and five female rats received 0.5 ml of Margosan-O by intraperitoneal injection. The remaining six (control) rats were left untreated. The rats were maintained until the 14th day after substance administration. At that time, they were again anesthetized and weighed and a blood sample was taken as on day 0. Blood samples were submitted to repeat the analyses conducted at the study initiation, i.e., complete blood counts with differential and serum protein electrophoresis.

Body weights on day 0 and day 14 were combined for each of the four groups and a mean and standard deviation were calculated. All surviving rats gained weight and appeared active and healthy. No differences were evident between test and control animals. There were no significant changes in the hematology of the treated rats between day 0 and day 14, although statistical analysis of the electrophoretic pattern showed differences ($p < 0.05$) in the globulin fractions during that period. The differential count showed a statistically significant change in the polymorphonuclear count, but none of the other differential counts differed statistically.

Comparison of the changes in blood values of the control rats over 14 days with those in the treated rats did not uncover significant differences among any of the parameters measured. This suggests that the changes noted above were not treatment-related, but rather that they were normal and not unusual. The results of this study suggest that the test material does not cause an adverse immune response.

Test 11, Sensitization. This test was done on guinea pigs, which were shaved and patched with Margosan-O test material for 6 h. The procedure was repeated on alternate days for a total of nine applications. A retest dose was applied after 14 d with duplicate patches, and the reaction was read 24 h later. Margosan-O does not produce sensitization.

Test 12, Mutagenicity. This is the traditional Ames mutagenicity study used in the U.S. on five strains of *Salmonella typhimurium*. Results of this test indicate that the Margosan-O concentrate is nonmutagenic.

Additional Test: this test was done voluntarily and was not ordered by the EPA. With the assistance of the University of California Apiary at Riverside, CA, Vikwood Ltd. ordered a "Bee Adult Toxicity Test" on honeybee worker adults. Margosan-O was administered as a direct contact chemical using field dosages up to 4478 ppm a.i./ha. It was found to be benign to honeybees at well above the recommended dosage of 20 ppm (diluted, as a foliar spray) for a common pest, the gypsy moth, *Lymantria dispar* (L.).

VI. FUTURE GOALS

1. Continued cooperation with the USDA research laboratories at Beltsville, MD and other IR4 stations to determine expanded uses of Margosan-O, or other derivative products from the neem seed.
2. Procure seeds from newly discovered areas of known stands of neem trees, and assay them for azadirachtin content. The intent would be to source new stock for cultivars of high quality.
3. Improve extractive procedures to obtain the highest yield at the lowest extraction costs and to increase the market share of an efficacious, economical, and safe pesticide, Margosan-O.
4. Establish neem plantations in various tropical arid countries with the goal of sharing the neem seed harvest and technology for extraction.
5. Establish joint venture extraction units in other countries for the development of reliable sources of neem extract to further enhance Margosan-O by Vikwood Botanicals, Inc.
6. License established capable extraction companies outside the U.S. to manufacture Margosan-O and derivative neem products in order to supply distant markets without interruption.
7. Develop Margosan (free of oil for more efficient systemic action) and Margosan-D (an azadirachtin-rich formulated dust).
8. Develop an inexpensive nitrogen-rich fertilizer from residual neem cake with the addition of urea pellets to prepare an efficient fertilizer and nematicide. This would utilize the entire seed and hull, leaving no waste.
9. Through methodology developed in the USDA Beltsville laboratory, to develop a cockroach bait using Margosan-O to disrupt hormonal functions and eradicate even large colonies rapidly. Margosan-O is a proven attractant,[9] has shown no stomach poisoning effect, and is considered safe for use around children and pets in Eastern countries.
10. Attempt to complete, through sponsorship or joint venture, the 90-d dietary studies required by the EPA to register Margosan-O for use on food crops, or in the home as a cockroach bait. Tests are contemplated using both rats and pregnant rabbits and are estimated to cost in excess of $100,000.
11. Explore the possibilities of finding synergists to improve the efficiency of Margosan-O.

VII. UNITED STATES PATENT (19)

Larson
(11) Patent Number: 4,556,562
(45) Date of Patent: December 3, 1985

(54) Stable Anti-Pest Neem Seed Extract
(75) Inventor: Robert O. Larson, Sheboygan, WI
(73) Assignee: Vikwood, Ltd., Sheboygan, WI
(21) Appl. No.: 590,808
(22) Filed: March 19, 1984
(51) Int. Cl.⁴ A01N 65/00; A01N 43/16
(52) U.S. Cl 424/195.1; 514/453
(58) Field of Search 424/195, 279, 195.1, 424/453

(56) References Cited

U.S. PATENT DOCUMENTS

963,932 7/1910 Olsson.......... 424/127
3,663,253 5/1972 Stone........... 106/204

Other Publications

Radwanski, S. A. et al., Vegetative fallows and potential value of the neem tree (*Azadirachta indica*) in the tropics, *Econ. Bot.,* 35(4), 398, 1981.
Stokes, J. B. et al., Effect of sunlight on azadirachtin: antifeeding potency, *J. Environ. Sci. Health,* A17(1), 57, 1982.
Uebel, E. C. et al., Preparative reversed-phase liquid chromatographic isolation of azadirachtin from neem kernels, *J. Liq. Chromatogr.,* 2(6) 875, 1979.
Warthen, J. D. et al., *J. Liq. Chromatogr.,* 7(3), 591, 1984.
Primary Examiner — Allen J. Robinson
Assistant Examiner — John W. Rollins

(57) ABSTRACT

Storage stable composition effective as a biorational agent for protection against pest, e.g., Japanese beetles, is a diluted ethanol neem seed extract comprising from about 2000 to about 4000 ppm azadirachtin and having a pH ranging from about 3.5 to about 6.0.

18 Claims, No Drawings

UNITED STATES PATENT AND TRADEMARK OFFICE
CERTIFICATE OF CORRECTION

Patent No.: 4,556,562
Dated: December 3, 1985
Inventor: Robert O. Larson

It is certified that error appears in the above-identified patent and that said Letters Patent is hereby corrected as shown below:

In the list of OTHER PUBLICATIONS on the cover page, additionally include the following:

Recent Advances in Phytochemistry, Chapter 11, by Koji Nakanishi, 1979, pp 283, 288, 289.

Chemical Abstracts 102:41598u, 1985, "Isolation from Neem Oil of Active Principles Evincing Oviposition Deterrent Activity in Insects"

Claim 10, line 3, after "3,500" insert — ppm azadirachtin — .

Claim 16, line 2, "12" should be — 15 — .

Stable Anti-Pest Neem Seed Extract
Technical Field

This invention relates to a composition for protecting food and fiber crops from harmful pests, and also to a process for preparing the same.

Background of the Invention

For many years now, several powerful and effective insecticides have been used to protect food and fiber crops. More recently, there has been a great deal of controversy about the effect of these on the environment and some of the insecticides which have been in common use have been banned. Furthermore, other insecticides which are still in use are considered to be potentially harmful to the environment but are required to be used for lack of other alternatives.

As a result, a search has been going on for "biorational pesticides". These are compositions which would deter insects or other pests but would have no or minimal harmful effect on the environment.

One agent known to protect crops from pests is azadirachtin which is a natural product found in the seeds of the neem tree (*Azadirachta indica* A. Juss). The neem tree is found in India, Pakistan, Bangladesh, Burma, Thailand, Malaysia, and Africa, for example.

Azadirachtin has been extracted from neem seeds and found to have antifeedant (deters insects from feeding on plants) and growth regulation potency against several pests including Japanese beetles, fall armyworms, locusts, termites, grasshoppers, tobacco hornworms, tobacco budworms, caterpillars, gypsy moths, rice weevils, aphids, cotton boll moths, and many others. It is readily applied by coating seeds or by applying a spray to the crops themselves. See for example *J. Environ. Sci. Health,* A17 (1), 57, 1982 by J. B. Stokes and R. E. Redfern of the USDA.

While azadirachtin is a known agent, it has not come into commercial use because it has stability problems. For example, its instability in sunlight has been known and is subject of the aforementioned article by Stokes and Redfern. That article indicates that sunlight degradation is hindered by leaving some neem oil in with the azadirachtin or by adding other plant oil and that further aid is given by including a sunscreen (an ultraviolet absorbing additive). Azadirachtin has also been found to have storage stability problems whereby it deactivates in the container or after application simply on the passage of time. No solution has been suggested in the published literature for these problems.

Summary of the Invention

It has now been discovered that a storage-stable composition containing azadirachtin comprises from about 2000 to about 4000 ppm azadirachtin and has a pH ranging from about 3.5 to about 6.0. By storage-stable is meant retaining more than about 80% of its potency when in the form of an emulsion 8 weeks or more after formation of the emulsion and more than 50% of its potency even after 2 years.

The composition, in the preferred embodiment, is a diluted ethanol extract of neem seeds wherein the pH has been adjusted to the aforestated level. This composition besides being storage-stable, has the advantage of being economical to prepare. It is suitable for use as a biorational agent for protection against pests.

Preferably, the composition comprises from about 20% to about 25% neem oil. Neem oil is the triglyceride constituent normally found in the neem seeds. As indicated in the aforementioned article, it has an effect on protecting the azadirachtin from sunlight degradation. It is readily extracted from the neem seeds at the same time as the azadirachtin.

A process herein comprises the steps of

1.   Forming neem seed particles
2.   Extracting the particles with extraction agent compromising ethanol at a temperature ranging from about 60°C to about 90°C and separating the extract to obtain a solution comprising from about 5000 to about 10000 ppm azadirachtin
3.   Diluting to form an emulsion comprising from about 2000 to about 4000 azadirachtin

4.   Measuring the pH, and if it is outside the range of from about 3.5 to about 6.0, adjusting the pH to range from about 3.5 to about 6.0

All percentages and parts herein are by weight unless otherwise stated.

Detailed Description

The composition herein preferably contains azadirachtin at a level ranging from about 2500 to about 3500 and has a pH ranging from about 3.8 to about 4.2. Optimally, it contains azadirachtin at a level of about 3000 ppm and has a pH ranging from about 3.84 to about 4.0.

It has been found that the concentration of azadirachtin and pH level are both very important in respect to storage stability and that inside the limits recited herein azadirachtin levels are substantially maintained up to 8 weeks or more (with up to 65% potency being retained even after 2 years); whereas outside the ranges herein, azadirachtin levels are reduced for example in a shorter period of time, for example in about 3 weeks or even less.

As previously indicated, it is highly desirable to retain at least some of the neem oil present in the neem seeds in the composition herein. It is noted that the amount of neem oil present in neem seeds varies substantially from batch to batch and according to the literature comprises from about 20% to about 50% of neem seeds.

It is highly preferred to include an emulsifying agent in the composition so that the azadirachtin and neem oil and any other ingredients are kept uniformly distributed in the composition. The percentage of the composition which is emulsifier normally depends on the emulsifier which is used. Typically, the emulsifying agent (i.e., the active ingredient) is used in an amount ranging from about 0.2% to about 30% of the composition and often is used at a level of from about 15% to about 25%.

Preferred emulsifying agents are those normally utilized in foods and include sorbitan esters, ethoxylated and propoxylated mono- or diglycerides, acetylated mono- or diglycerides, lactylated mono- or diglycerides, citric acid esters of mono- or diglycerides, sugar esters, polysorbates and polyglycerol esters. Preferred emulsifier is polyoxyethylene sorbitan monolaurate which is sold under the name Tween® 20.

It is preferred to include as an optional ingredient a sunscreen at a level, for example, of about 1% to about 2.5%. A preferred sunscreen is *p*-aminobenzoic acid or its esters.

We turn now to the process herein for economically making a preferred storage-stable composition herein.

The step of forming neem seed particles involves comminuting or otherwise size-reducing, typically to form particles ranging in size from about 350 μm to about 2 mm (maximum dimension). A very preferred size is that of the size of a regular grind of coffee. According to the U.S. Department of Commerce (see Coffee Brewing Workshop Manual, p. 33, published by the Coffee Brewing Center of the Pan American Coffee Bureau), regular grind has a particle size defined by 33% retained on 14-mesh Tyler standard sieve, 55% retained on a 28-mesh Tyler standard sieve and 12% passing through a 28-mesh Tyler standard sieve. The comminuting or size-reduction is readily carried out on standard apparatus for this purpose, for example a standard commercial feed mill or a coffee mill such as an American Duplex Model 480 grinder.

The "regular grind" is preferred because extraction is easily carried out on this particle size. The avoidance of a high percentage, for example 30% or more, of fine particles is preferred because the inclusion of such a large percentage of fine particles will require extraction by special means and hinder separation of the extract from the particles.

If desired, the particles may be compressed, for example, using a roll mill to disrupt cellular structure for easier extraction.

The extraction is preferably carried out using ethanol at a temperature ranging from about 75°C to about 85°C and optimally at 80°C. Very preferably, extraction is carried out to obtain a solution comprising about 10,000 ppm azadirachtin which is the solubility limit for azadirachtin in ethanol. Very preferably extraction is carried out to obtain a solution containing from about 40% to about 45% neem oil.

The extraction is readily carried out by passing a batch of the extracting agent which has been heated to the aforementioned temperature through the particles of neem seeds. The separation of the extract from the particles is readily carried out for example using vacuum filtration or centrifugation or other common separating method.

The extraction is preferably carried out by a process involving multiple passes of the extracting agent through the particles. This involves, for example, passing a batch of extracting agent through the particles and recovering it and then passing the same batch of extracting agent through the particles again and continuing this process until a total of 8 to 12, preferably 10, passes of the extracting agent have been made through the particles, with the extracting agent being heated to the aforementioned temperatures between passes if necessary.

The step of diluting to form an emulsion comprises adding diluent, preferably water, and emulsifying agent to obtain an emulsion comprising from about 2000 to about 4000 ppm azadirachtin and usually from about 20% to about 25% neem oil. Where the emulsifying agent is obtained commercially in very diluted form, this diluting step is carried out simply by admixing the emulsifying agent as purchased with the extract. Preferably the emulsifying composition contains water in addition to the emulsifying agent and the water forms at least the main portion of the diluent. The diluting step, of course, should be carried out prior to any substantial degradation of the azadirachtin potency.

The pH of the resulting diluted extract varies from batch to batch of neem seed starting material. In a very few instances the pH may already be within the required limits. However, usually pH adjustment is required. Even if pH measurement indicates a pH within the aforestated limits, optimization may be desired. The pH adjustment is carried out by the addition preferably of a base which is not toxic or harmful to the environment in the concentration utilized. The preferred agent is ammonium hydroxide as the ammonium constituent is nutritious to the plants to which the composition is applied. Other pH adjustment agents include, for example, sodium or potassium hydroxide.

The sunscreen is preferably added after pH adjustment but it can be admixed before this.

Preferred embodiments of the invention herein are illustrated in the following specific examples.

EXAMPLE I

40 lb of neem seeds imported from India is ground using a standard commercial feed mill to the size of a regular grind of coffee.

250 g of the grind is positioned in a columnar extraction system and 250 ml of ethanol at 80°C is passed through the grind using vacuum filtration to recover the extract. The product is heated up to 80°C again and is passed through the grind again. This is repeated for ten cycles, i.e., the ethanol is passed through the grind a total of ten times with the concentration of extracted matter increasing each time. At the end of the ten cycles, the extract contained 10,000 ppm azadirachtin and 40% neem oil. About 90% of the available azadirachtin is recovered from the seeds. Some ethanol flashes off during the processing.

To the extract is added Tween® 20 (consisting of water and polyoxyethylene sorbitan monolaurate) to obtain a diluted extract containing 3000 ppm azadirachtin and 20% neem oil.

To the resulting emulsion is added ammonium hydroxide to adjust the pH to 4.0.

The resulting product is an effective biorational agent against a wide spectrum of pests and is shelf stable for more than 8 weeks after formulation.

Product containing 3000 ppm azadirachtin and having a pH of 4.0 maintained under normal shelf life conditions (without refrigeration or addition of sunscreen) was found to have retained 65% of its potency even 2 years later.

EXAMPLE II

Processing is carried out as in Example I except that the emulsifying agent is Triton® X-100 (0.5 parts to one part concentrated extract) and neutralization is carried out to pHs as listed below. Triton® X-100 is an octyl phenoxyl polyethoxy ethanol.

Azadirachtin levels (ppm) on storage were measured with the following results:

AZADIRACHTIN LEVELS (ppm) ON STORAGE AT SEVERAL pH VALUES

	Storage Time (weeks)				
pH	0	1	2	3	8
3.84	3138	2841	3214	2500	3200
4.80	3047	3105	2680	2460	2900
5.99	2473	2847	3143	2260	2370
6.77	2644	2408	2877	2180	—
7.90	2690	2010	1431	1520	—

As can be seen from the above tables, solubility of azadirachtin is reduced as pH is raised and there is a storage stability advantage at lowest three pHs.

EXAMPLE III

Processing is carried out as in Example I except that extraction and separation of extract from neem seed particles is carried out by centrifuging. The resulting product is effective against a wide spectrum of pests.

REFERENCE EXAMPLE I

Product containing 10,000 ppm azadirachtin and having a pH less than 3.5 was found to have its azadirachtin concentration reduced to 2500 ppm within approximately 1 month.

While the foregoing describes preferred embodiments, modifications within the scope of the invention will be evident to those skilled in the art.

For example, some of the ethanol in the extracting agent, e.g., 20% by volume, can be replaced with water.

Thus, the scope of the invention is intended to be defined by the claims.

What is claimed is

1.   A process for preparing a storage stable neem seed extract, said process comprising the steps of:

a. Forming neem seed particles ranging in size from about 350 μm to about 2 mm (maximum dimension)
b. Extracting the particles with extracting agent comprising ethanol at a temperature ranging from about 60°C to about 90°C and separating the extract to obtain a solution comprising from about 5000 to about 10,000 ppm azadirachtin
c. Diluting to form an aqueous emulsion comprising from about 2000 to about 4000 ppm azadirachtin
d. Measuring the pH, and if it is outside the range of from about 3.5 to about 6.0, adjusting the pH to range from about 3.5 to about 6.0

2. Process as recited in claim 1 wherein in step (b), the solution obtained contains from about 40% to about 45% neem oil.
3. Process as recited in claim 2, wherein extracting is carried out by passing a batch of extracting agent through the particles 8 to 12 times and wherein the extracting agent is utilized at a temperature ranging from about 75°C to about 85°C.
4. Process as recited in claim 3, wherein diluting to form an emulsion is carried out by using water in combination with nonionic emulsifying agent.
5. Process as recited in claim 4, wherein the extracting is carried out utilizing extracting agent at about 80°C and passing such through the particles ten times to obtain a solution comprising about 10,000 ppm azadirachtin, and wherein diluting is carried out to form an emulsion comprising about 3000 ppm azadirachtin and wherein the pH is adjusted to about 4.
6. Process as recited in claim 5, wherein the pH adjustment is carried out by addition of ammonium hydroxide.
7. Process as recited in claim 6, wherein separation of extract from the particles is carried out utilizing vacuum filtration.
8. Process as recited in claim 6, wherein separation of extract from the particles is carried out utilizing centrifugation.
9. Process as recited in claim 6, wherein sunscreen comprising p-aminobenzoic acid or its esters is added to be present at a level ranging from about 1% to about 2.5%.
10. Process as recited in claim 1, wherein the diluting is carried out to form an emulsion comprising from about 2500 to about 3500 and the pH is adjusted to be from about 3.8 to about 4.2.
11. Process as recited in claim 10, wherein diluting is carried out to form an emulsion comprising about 3000 ppm azadirachtin.
12. A storage stable aqueous azadirachtin-containing composition, said composition comprising from about 2000 to about 4000 ppm azadirachtin, said composition having a pH ranging from about 3.5 to about 6.0.
13. A storage stable composition as recited in claim 12 comprising from about 20% to about 25% neem oil.
14. A storage stable composition as recited in claim 13 which is derived from an ethanol extract of neem seeds.
15. A storage stable composition as recited in claim 14 which comprises from about 2500 to about 3500 ppm azadirachtin and has a pH ranging from about 3.8 to about 4.2.
16. A storage stable composition as recited in claim 12, wherein the azadirachtin is present at a level of about 3000 ppm.
17. A storage stable composition as recited in claim 16 which contains from about 15% to about 25% nonionic emulsifying agent.
18. A storage stable composition as recited in claim 17, comprising additionally from about 1% to about 2.5% sunscreen comprising p-aminobenzoic acid or its esters.

ACKNOWLEDGMENTS

Vikwood Ltd. (the originator of Margosan-O®) and Vikwood Botanicals, Inc., the present holder of the EPA Registration No. 58370-1 and the U.S. Patent No. 4,556,562, are deeply indebted to the fine cooperation and assistance given, without stint, by the U.S. Department of Agriculture, Beltsville, MD and its energetic team of chemists and entomologists who provided guidance and assistance over the past 8 years.

REFERENCES

1. **Ketkar, C. M.,** Utilization of Neem and its By-Products, Report of the Modified Neem Cake Manurial Project 1969—76, Directorate of Non-Edible Oils and Soap Industry, Khadi and Village Industries Commission 3, Bombay, India, 4.

2. **Redknap, R. S.,** The use of crushed neem berries in the control of some insect pests in Gambia, in *Natural Pesticides from the Neem Tree (Azadirachta indica A. Juss),* Schmutterer, H. and Rembold, H., Eds., GTZ Press, Eschborn, West Germany, 1981, 213.

3. **Radwanski, S. A.,** Home made neem insecticides for the tropics, *Econ. Bot.,* 4, 4, 1982.

4. **Anon.,** Tobacco Mosaic — Its Control With Neem Leaf Decoction, Indian Leaf Tobacco Co. Ltd., Pamphlet No 24 RJY 6'73 3,000.

5. **Butterworth, J. H. and Morgan, E. D.,** Isolation of a substance that suppresses feeding in locusts, *Chem. Commun.,* 23, 1968.

6. **Bilton, J. N., Broughton, H. B., Ley, S. V., Lidert, Z., Morgan, E. D., Rzepa, H. S., and Sheppard, R. N.,** Structural reappraisal of the limonoid insect antifeedant azadirachtin, *J. Chem. Soc., Chem. Commun.,* 968, 1985.

7. **Kraus, W., Bokel, M., Klenk, A., and Poehnl, H.,** The structure of azadirachtin and 22,23-dihydro-23β-methoxyazadirachtin, *Tetrahedron Lett.,* 26, 6435, 1985.

8. Chemische Fabrik Carl Roth GmbH Co., No. NP86, Karlsruhe, West Germany, cf. **Van Beek, T. A. and De Groot, A.,** Terpenoid antifeedants of natural origin, *Rec. Trav. Chim. Pays-Bas,* 105, 513, 1986.

9. **Adler, V. E. and Uebel, E. C.,** Effects of a formulation of neem extract on six species of cockroaches (Orthoptera:Blaberidae, Blattidae and Blattellidae), *Phytoparasitica,* 13, 3, 1985.

INDEX

Mouse, 143—144, 146
Mungbean, stored, insects pests on, 102
Musca, 57, 117
Musca autumnalis, 116
Musca domestica, 114—116
Musca domestica nebulo, 114
Muscular effects, 136
Mutagenicity, 106, 140, 160
MV-678, 106

N

Natural reproduction of neem, 12
Neem cake, 82—83, 114
Neem oil, 13, 82—83
Neem products
  antifeedant activity, 101—103
  insect growth inhibition and, 103—106
  insect repellent action of, 100—101
  insect reproduction and, 103—106
  pharmacology of, 135—139
  preparation of, 70—71
  toxicology of, 139—145
  types of, 70—71
Neemrich-100, 144
Neem scale, see *Aspidiotus orientalis; Palvinaria maxima*
Nematodes, 80, 81, 141—143
Nephotettix virescens, 147
Neuroendocrine system, of insect larva, 59
Nimbandiol, 21, 27, 39
Nimbic acid, 25, 38, 140
Nimbidin, 81, 135—137, 139—140, 144—145
Nimbidinic acid, 135
Nimbidinin, 21, 25
Nimbin, 21, 24—25, 29, 37—38, 71, 81—82, 140
Nimbinene, 21, 26, 39, 81
Nimbinic acid, 25, 38
Nimbinin, 22, 24
Nimbiol, 20—21, 35
Nimbocenone, 35
Nimbocinol, 33
Nimbocinone, 21—22, 33
Nimbola, 136
Nimbolide, 21, 26, 39, 136, 140
Nimbolin A, 21, 25, 38, 125
Nimbolin B, 21, 38
Nimocimolide, 119
Nimocinol, 21
Nimolicinoic acid, 21, 27, 40
Nimolicinol, 21, 27, 39
Nimolinone, 21, 33, 42
Nm1, 71—72
Nubian goat, 144—145
Nurse crop, 4
Nursery care of neem, 8—10
Nutritive effects, 137—139

O

Oak, insect pests on, 80, 88—89

Odites atmopa, 14
Oidium, 15
Oil, neem, see Neem oil
Okra, 76, 80
Oleic acid, 20, 116
Olive oil, 116
Oncolytic effects, 137
Oncopeltus fasciatus, 59—61
Onion, insect pests on, 77, 91
Onion leaf miner, see *Liriomyza trifolii*
Ootheca bennigseni, 75, 82
Orange, insect pests on, 80
Orange striped oakworm, see *Anisota senatoria*
Oriental cabbage webworm, see *Hellula undalis*
Oriental fruit fly, see *Dacus dorsalis*
Ornamental plants, insect pests on, 88—93
Orthacris simulans, 14
Ostracods, 140—141
Ostrinia turnacalis, 58

P

Paddy, stored, insect pests on, 103—104, 106
Palli, 99
Palmitic acid, 20
Palm oil, 106
Palvinaria maxima, 14
Panonychus citri, 79
Papilio demodocus, 80
Parasites, 15, 122—123
Pasture, neem combined with, 6—7
Patent, United States, 161—167
Pea, stored, insect pests on, 102, 105
Pectinophora gossypiella, 58
Pediculus, 123
Pennisetum typhoideum, see Bajra
Pentanortriterpenoids, 20
Periplaneta americana, 123—124
Persian lilac extract, 122
pH, soil, neem growth and, 3
Pharmacology of neem products, 135—139
Phentoate, 121
Phoma jolyana, 15
Phycita melongenae, 77
Phyllocnistis citrella, 79—80
Phyllotreta, 73, 82
Phytoseiulus persimilis, 79
Pieris, 73
Pigeon pea, 75, 102
Pimples, 134
Piperonyl butoxide, 72
Planococcus citri, 79
Plantation characteristics, 3—4
Plasmodium berghei, 140
Plutella xylostella, 71—73, 81
Podagrica, 76, 82
Pod borer, see *Heliothis armigera; Maruca testulalis*
Pod fly, see *Melanagromyza obtusa*
Pod sucking bud, see *Acanthomia horrida*
Pole production, 5
Pongamia glabra, see Karanja oil

Printed and bound by CPI Group (UK) Ltd, Croydon, CR0 4YY

22/10/2024

01777630-0018